Floods in North West England: a history c. 1600–2008

by
Sarah Watkins & Ian Whyte

Centre for North-West Regional Studies
Lancaster University
2009
Series Editor: Jean Turnbull

Floods in North West England: a history c. 1600–2008

This volume is the 56th in a series of Occasional Papers published by the
Centre for North-West Regional Studies at the University of Lancaster

Text Copyright © Sarah Watkins & Ian Whyte 2009.
All rights reserved.
The moral rights of the authors have been asserted.

Designed, typeset, printed and bound by
4Word Ltd, Bristol

British Library Cataloguing in Publication Data. A catalogue record for this book is
available from the British Library

ISBN 978-1-86220-217-7

Contents

Acknowledgements

The authors would like to express their gratitude to the Leverhulme Trust whose funding, under grant F00 185L allowed the research on which this study is based. The research involved consulting material in libraries and archives from Carlisle to Manchester.

We would like to express our thanks to the staff of:

Carlisle Record Office
Carlisle Central Library
Penrith Library
The Armitt Library, Ambleside
Kendal Record Office
Kendal Reference Library
Lancaster Reference Library
Lancaster University Library
Fleetwood Library
Lancashire Record Office, Preston
Bolton Record Office
Harris Library, Preston
Salford Local History Library

for their help. Also to colleagues in the Lancaster Environment Centre at Lancaster University for stimulating discussions and ideas, especially Dr. Harriet Orr, and to Simon Chew for drawing the maps. We would particularly like to thank Professor Keith Beven for reading a draft of the book and for providing many valuable ideas and suggestions.

The publisher would like to thank the individuals and organisations who gave permission for the Centre to use photographs in this book. Every effort has been made to trace copyright holders of images, but this has not always been possible. We would be pleased to hear from any copyright holder who we have not been able to identify.

The publisher gratefully acknowledges the assistance of the Environment Agency, although the opinions expressed in this publication are those of the authors and are not necessarily shared by the Environment Agency.

List of Illustrations

List of Tables

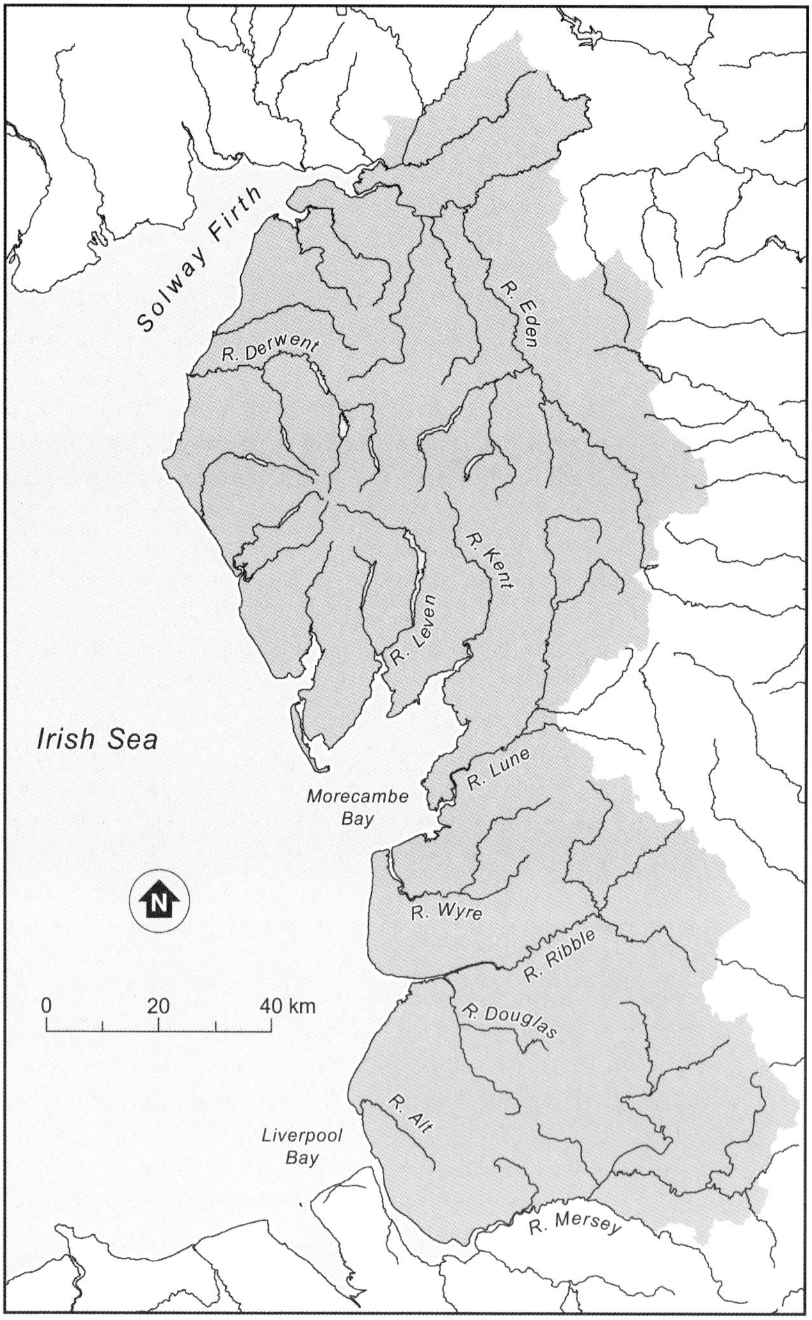

Solway Firth
R. Derwent
R. Eden
R. Kent
R. Leven
Irish Sea
Morecambe Bay
R. Lune
R. Wyre
R. Ribble
N
0
20
40 km
R. Douglas
R. Alt
Liverpool Bay
R. Mersey

1: North West River Catchments. Map drawn by Simon Chew, Lancaster Environment Centre.

CHAPTER 1

Introduction

On the 8–9 January 2005 a month's rain fell within 36 hours on the Eden valley in Cumbria, accompanied by winds gusting to over 100 miles an hour. The result was flooding in the city of Carlisle on a scale generally considered to have been even more severe that the previous record level reached in 1822, when the water had been a metre lower at the Eden Bridge. Some of the city's flood defences were successful in protecting around 6,000 people but the homes of thousands more were overwhelmed by the extreme flow of water. Most of the flood barriers in the city were only designed to cope with a level of water which was likely to occur once every 70 years. The flood left three people dead and 150 people were admitted to hospital. The police and fire stations were inundated, and the Civic Centre flooded. Over 1,700 properties in the city were damaged by water and nearly 300 businesses were badly affected. Around 10,000 people had to be evacuated from their homes, at least 15 of them by helicopter (Figures 2 & 3). The overall cost of repairing the damage ran to nearly £300 million but some of the effects, such as ill health and stress, were longer term and are hard to assess. As well as Carlisle, smaller numbers of householders were affected by flooding in towns and villages throughout Cumbria including Appleby, Kendal, Keswick, and Shap. Power was lost in 76,000 homes when an electricity sub-station was swamped. Thirty months later, in the summer of 2007, some people had still not been able to return to their homes. It was no consolation to the flood victims that, on past trends, there is not likely to be another flood of this magnitude for perhaps another century and a half.

The flooding of Carlisle was only one of a cluster of flood disasters at a range of scales which have hit the headlines in recent years. The flooding of New Orleans in August 2005, as a result of the breaching of the levees by a storm surge caused by Hurricane Katrina, made us realise how vulnerable large cities can be, even in the wealthiest nation on earth. Floods on the Danube and other central European rivers in 2002 caused severe damage to historic buildings and art treasures in cities like Prague. Closer to home the sudden flash flood at Boscastle in Cornwall in August 2004 brought unsettling echoes of the disaster at Lynmouth in Devon in 1952 when 24 people were drowned and 420 made homeless. The floods

2: Flooding in Carlisle, January 2005. The worst for over 150 years. The fire station and the police station. Reproduced by permission of David Ramshaw.

in Yorkshire, the West Country and the Midlands in June and July 2007, affecting tens of thousands of homes and businesses with costs estimated at more than £3 billion, made us realise that widespread floods in Britain can happen in summer as well as winter. But it was the flooding of Carlisle especially which served as a reminder that even in Britain, a country where environmental extremes are relatively rare, flooding is still a danger to be reckoned with. The Environment Agency, which monitors flood hazards, has concluded that at least five per cent of the population of Britain live in flood-prone areas and that two million homes are at risk. Despite rising expenditure on flood protection schemes it remains true that for people living and working in the north west floods are one of the main environmental hazards which they are likely to experience directly at some point in their lifetimes as householders, workers or travellers. The recent events mentioned above have emphasised the need for better understanding and management of flood hazards, including tighter controls on new building developments on river floodplains.

Some past flood disasters are also firmly embedded in people's memories. Lynmouth, mentioned above, and the east coast of England sea floods in 1953, are still used as benchmarks against which more recent catastrophes are measured. Floods which happened before the mid-twentieth century, however, are recorded mainly as footnotes in local histories. Nevertheless, though almost forgotten, they often devastated

3: Flooding in Carlisle, January 2005. The view towards the castle. Reproduced by permission of David Ramshaw.

communities, such as the sea flood which swamped Fleetwood in 1927 or the river floods in 1771 which affected the north west badly and had an even greater impact east of the Pennines, destroying every bridge but one on the River Tyne. It has been claimed that this was the worst river disaster in Britain in the last 1,000 years.[1]

There are concerns about the possibility of increased flooding in the future. An extreme scenario suggests a possible sea level rise of up to seven metres by 2100, due to the melting of the Greenland ice sheet, highlighting the threat to low-lying coastal areas in Britain and elsewhere. There has already been a 20 per cent increase in average winter rainfall in the Lake District over the last 70 years.[2] There are also suggestions that over the next few decades there will be less rainfall in summer in north-west England but perhaps 20 per cent more in winter, occurring in heavier bursts and, potentially, generating more severe floods.[3]

Today floods are often blamed on local authorities or government agencies which, it is claimed, have not provided adequate defences. In earlier times floods were more often seen as acts of God, purely natural phenomena against which little could be done. Often, however, today, as in the past, the impact of floods can be worsened directly and indirectly by human activities. Interference with the flow of rivers by erecting barriers such as bridges, embankments and weirs are obvious examples. The reclamation of low-lying coastal areas for agriculture leaves them vulnerable to the overtopping of defences by the sea during tidal surges. Building new houses in riverside locations is simply asking for trouble.

The north west has areas with the highest rainfall totals in England, streams with the steepest gradients in the country and estuaries and coasts with some of the highest tidal ranges in Europe.[4] Cumbrians and Lancastrians ought to be good at coping with the problems of excess water – but are they? Communities within the region still remain vulnerable to flooding despite, and sometimes because of, modern technology. There is increasing concern that people in Britain have become more rather than less susceptible to flooding in recent times compared with the past, not merely due to climatic shifts, but to direct and indirect human agency. The progressive colonisation of floodplains, by factories since the start of the Industrial Revolution and by housing during the late nineteenth and twentieth centuries, accounts for part of the problem. There has been concern at both local and national levels that planning permission is being granted too readily for residential developments on sites which are known to be vulnerable to flooding. In recent years the Environment Agency has objected to many proposals for housing developments in high flood risk areas, but around a fifth of these have nevertheless been approved. It has also generally been the case until recently that developers have not always been required to contribute adequately to the cost of flood protection for the sites they build on (see Chapter 9).

In addition to this, land use changes may have had an impact on the speed and intensity of flooding for a given volume of rainfall. The spread of impermeable street and road surfaces with efficient drains, the extension of built-up areas, and the effects of the drainage of agricultural land all send rain into streams and rivers more quickly and limit the ability of the ground to absorb the water.[5] Even the modern trend of paving front gardens to provide off-road parking space has been blamed as a contributory cause of flooding by increasing the speed at which rainwater drains off.

To assess more effectively the risks and likely impacts of flooding in the future it is important to have a firm grasp of the chronology, causes and impacts of flooding in the past. Because of their very localised extent most floods are soon forgotten and there is a lack of information on how frequent and damaging they have been in the past. As we will see in Chapter 3 there are many ways of identifying past floods.

The dating of river sediments using radiocarbon analysis has suggested that there were periods within historic times which experienced relatively frequent or few major floods. The first of these was the medieval warm phase (*c.* AD 1100–1300), when rainfall increased by around three per cent.[6] The second was the final phase of the post-medieval cold spell widely known as the Little Ice Age. Between about AD 1700 and 1850 there was a marked increase in autumn and winter floods.[7] The period before this, spanning the coldest phase of the Little Ice Age which seems to have reached its worst in Britain at the end of the seventeenth century, was distinguished by less rainfall, with annual totals down by about seven and ten per cent though there may have been more snowfall.[8]

Research into the occurrence and scale of floods in the past has tended to ignore the potential of historical records. However, a good deal of historical information exists to extend flood chronologies back to at least the start of the seventeenth century. Previous studies of historical records as a source of data on floods include those of Archer for floods on the rivers of north eastern England.[9] Lindsey McEwen has established chronologies of floods for various Scottish rivers including the Tweed, Tay and Dee.[10] A study of flood chronologies on the rivers Severn and Wye identified a period of more frequent flooding between the 1840s and the 1880s, another from the 1940s onwards, with a phase of reduced flooding in between.[11] There have also been some detailed studied of flood histories in parts of Ireland and Wales.[12] However, for the north west less information is available. The Environment Agency's various catchment flood management plans, which are available online from their website, highlight some notable past floods but often only as far back as the early twentieth century.[13]

To assess the scale of floods which may recur at intervals longer than once a century requires the study of historical records because data from river flow gauges go back only 30 years or so. Detailed knowledge of the chronology, causes and impacts of floods on a timescale of centuries is not only of historical interest but is also of practical relevance in planning for new developments and land use changes. It provides a historical context for current flood problems and may help to answer key questions relating to present flood risks such as: are extreme floods becoming more frequent and severe?[14]

But there is also an important human dimension to the impact of flooding. People's perception of flood hazards seem to have varied through time due to economic, social and cultural changes. Farming communities in the past may have been better prepared for, and more resilient to, the effects of flooding than modern societies. Attitudes to flooding and demands for flood protection are likely to have been affected by religious beliefs, variations in material wealth, and changing perceptions of authority. The increase in the range and cost of household possessions, including fitted carpets and electrical goods, since the Second World War is likely to have influenced people's perceptions. The modern 'compensation culture' attitude that views every natural disaster as somebody's fault, something that should be paid for by someone else, did not occur in the past. The costs of modern floods have been greatly increased by the vulnerability of electric and electronic equipment in people's homes to even limited, short-term exposure to water. Insurance policies which undertake to replace damaged goods with new ones have a similar effect. The ways in which the impact of floods have been presented by the media have also varied through time with a move away from reporting their effects on the countryside towards emphasising damage to urban property from the eighteenth to the twentieth centuries in line with a shift in the balance of population from rural to urban and a decline in the importance of farming within the economy.

The aim of this book is to reconstruct the history of flooding and its effects in north west England from the Mersey to the Solway. Chapter 2 reviews the environmental background to flooding, explaining some of the approaches and terminology used by hydrologists in studying floods. Chapter 3 looks at a range of historical sources which provide information on past floods and also reviews some of the evidence for long-term variations in flooding within the region. Chapter 4 examines the impact of flooding on communities within the region. Chapter 5 considers the extent to which human activity has worsened flooding. Chapter 6 presents the flood histories of two north west river basins: the predominantly rural River Eden in Cumbria and the heavily urbanised River Mersey and Irwell in Greater Manchester. Chapter 7 looks at the particular problems of flash floods in small catchments while Chapter 8 chronicles coastal flooding. Chapter 9 examines the history of various voluntary flood relief schemes designed to help victims. Finally in Chapter 10 we consider attempts to control flooding in the past, at the present time and in the future.

Notes

[1] D. Archer, *Land of Singing Waters. Rivers and Great Floods of Northumbria* (Spedden Press, 1992).

[2] H. G. Orr, The Impact of Recent Changes in Land Use and Climate on the River Lune: Implications for Catchment Management. Unpub. PhD Thesis, Lancaster University (2000).

[3] J. Mayes, 'Changing regional climatic gradients in the UK', *Geographical Journal* 16 (2) (2000), pp. 125–38.

[4] P. A. Barker, R. L. Wilby & J. Borrows, 'A 200–year precipitation index for the central English Lake District', *Hydrological Sciences Journal* 49 (5) (2004), pp. 769–85.

[5] M. Robinson & K. J. Beven, 'The effect of mole drainage on the hydrological response of a swelling clay soil', *Journal of Hydrology* 64 (1983), pp. 205–23.

[6] D. Higgitt & E. M. Lee, *Geomorphological Processes and Landscape Change. Britain in the last 1000 years* (Blackwell, 2001), p. 17.

[7] Higgitt & Lee, *Geomorphological processes*, p. 9.

[8] Higgitt & Lee, *Geomorphological processes*, pp. 9, 17.

[9] Archer, *Land of Singing Waters.*

[10] L. J. McEwen, 'The establishment of a historical flood chronology for the River Tweed catchment, Berwickshire, Scotland', *Scottish Geographical Magazine* 106, (1990), pp. 37–48; L. J. McEwen, 'The use of long-term rainfall records for augmenting historic flood series: a case study from the upper Dee, Aberdeenshire', *Transactions Royal Society of Edinburgh, Earth Sciences*, 78 (1987), pp. 279–85.

[11] G. M. Howe, H. O. Slaymaker & D. M. Harding, 'Some aspects of the flood hydrology of the upper catchments of the Severn and the Wye', *Transactions Institute of British Geographers* 41 (1967), pp. 33–58.

[12] N. L. Betts, 'The Antrim floods of October 1990', *Irish Geography* 25 (1992), pp. 138–45; D. B. Prior & N. L. Betts, 'Flooding in Belfast', *Irish Geography* 7 (1974), pp. 1–18; J. G. Tyrrel & K. J. Hickey, 'A flood chronology for Cork city and its climatological background', *Irish Geography* 24 (1991), pp. 81–90; R. P. D. Walsh, R. N. Hudson & K. A. Howells, 'Changes in the magnitude and frequency of flooding and heavy rainfall in the Swansea valley since 1875', *Cambria* 9 (1983), pp. 36–60.

[13] Environment Agency, *The Eden Catchment Flood Management Plan. Draft Scoping Report* (2005); *River Douglas Catchment Flood Management Plan. Draft Scoping Document* (2005); *River Ribble Catchment Flood Management Plan. Scoping Report* (2005).

[14] M. G. Macklin & B. J. Rumsby, 'Changing climate and extreme floods in the British uplands', *Transactions Institute of British Geographers* (New Series) 32 (2) (2007), pp. 168–9; A. Robson, 'Evidence for trends in UK flooding', *Phil. Trans. Roy. Soc. London*, A360 (2002), pp. 1327–53.

CHAPTER 2

The Environmental Background

To understand why floods occur it is necessary to know a little about hydrology (the study of water) and rivers. Rivers drain areas known as drainage basins or catchments, which are divided from each other by watersheds. Within the north west catchments range from the Mersey in the south to the Eden in the north. The length of a river or the area of its drainage basin can be used as measures of size but so can the discharge, the volume of water passing through a river's or stream's cross section at a specific location in a given period of time. This is usually measured in cubic metres per second, often abbreviated as 'cumecs'. While the area of a drainage basin and the length of a river will not vary significantly on a timescale of a few years, the discharge can change dramatically between summer drought conditions and winter floods. Variations in discharge over time at a particular site can be plotted as a hydrograph; ones relating to periods of flooding are known as flood hydrographs. These show how a river's discharge responds to heavy rainfall with a tendency for the graph to rise steeply to a flood peak and then fall more gradually, the flood peak occurring some time after the heaviest rainfall because of the time lag caused by the water finding its way into the stream channel. Continuous measurements of stream discharge are taken automatically by flow gauges maintained by the Environment Agency, the organisation responsible for implementing the government's environmental responsibilities, including flood defence and warning (Figure 4). The Environment Agency's website (http://www.environment-agency.gov.uk/subjects/flood/?lang=_e) provides detailed maps of the flood risk for every postcode in the north west and the rest of the country. The analysis of the Agency's streamflow records is an important source of information about flood hazards. Nearly 90 sites in north west England are regularly monitored. However, flow gauges may be unable to cope with the volume of extremely high floods. Even worse, regular discharge data from flow gauges in Britain usually only go back for a maximum of 50 years or so, and often much less, making it hard to assess the scale of flooding which may occur only once in a century or longer.

Within the north west, river basins have the same general characteristics as you trace them from their sources to the sea. The long profile of a river from source to mouth (its height above sea level plotted

4: Automatic
streamflow gauge,
Swindale, Lake
District. Photograph
Ian Whyte.

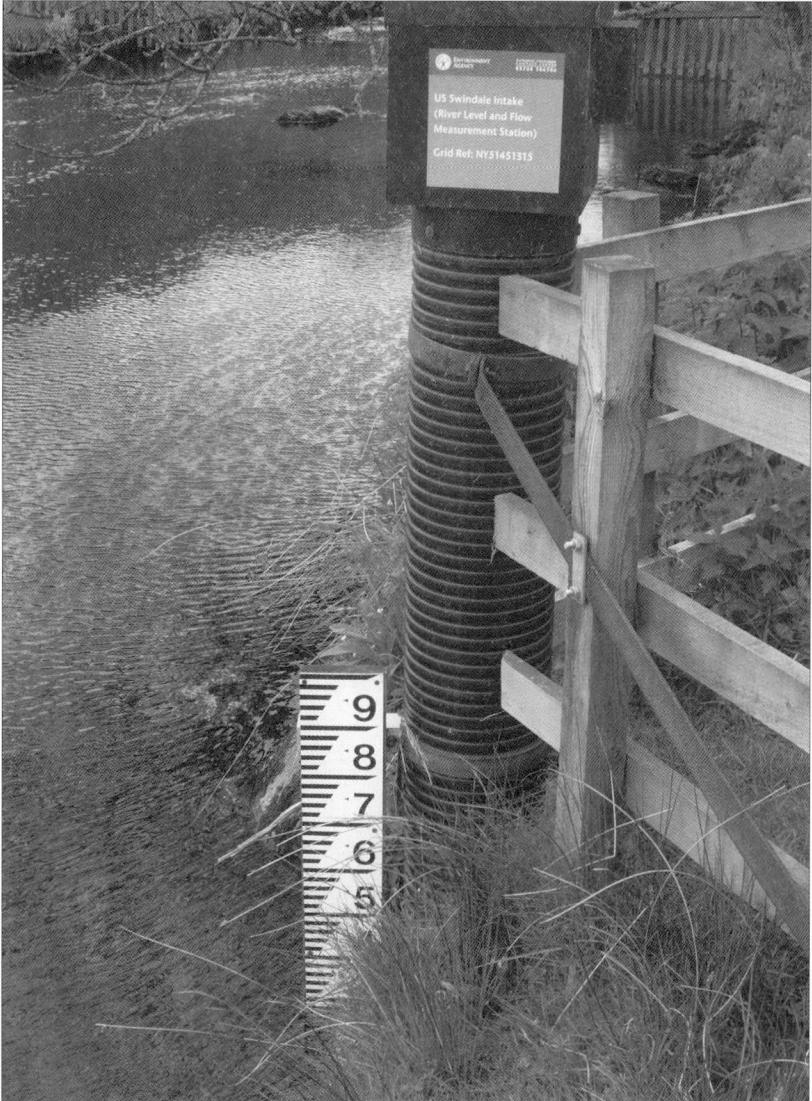

against its length) slopes steeply downwards near its source and gradually becomes more gentle: it can be divided into three distinct parts. In their upland or headwater sections (Figure 5) the gradients of streams (the amount of height they lose over a given distance) are marked, valleys are narrow and their sides often steep. Although waterfalls and rapids may make it look as though upland streams are flowing fast, in fact the friction from their rough and irregular beds slows down the speed of the current. Upland streams often have bedrock exposed in their channels. In upland source areas rainfall is high and the upland sections of streams are particularly vulnerable to severe flash floods caused by intense, though often very localised and short-lived, summer thunderstorms. Floods like

5: The upland section of a north west stream: Swindale. Photograph Ian. Whyte.

this can change the landscape more in an hour than in decades or centuries of less high streamflow, by cutting gullies in hillsides, causing peat slides, moving improbably large boulders and depositing sediments as fans of alluvium and gravel further down the valley (Figure 6). Some of the worst floods recorded in the region have been of this type (Chapter 7). Not surprisingly, the erosion of material rather than its deposition tends to be characteristic of upland river catchments.

6: A large fan of debris brought down by floods at the head of Chapel Beck, Howgill Fells. Photograph Ian Whyte.

In their middle sections, having left the hills behind, north west rivers flow across floodplains, level areas on either side which may be covered with water during a flood and where the alluvium that they have dropped usually buries the bedrock deeply. Rivers tend to flow in large loops, or meanders, as their gradients lessen. Rivers are wider than in the uplands and crossing them requires bridges or fords. The gradients of streams in their middle courses are gentler than in the uplands but the actual speed of the current increases because the channels are smoother with fewer obstructions to cause friction. During periods of heavy rainfall the level of water in such streams can build up until it completely fills the channel between the river banks, when it is said to be at bankfull stage, a situation which occurs on average every 18 months or so. If the river continues to rise beyond this, water will start to spill out over the floodplain. This small-scale flooding occurs frequently and usually causes little damage so that it may not even be reported or recorded as a flood. Some rivers may have their flows reduced by reservoirs which store water and reduce or prevent major floods. Other streams or even sections of the same river may experience increased flows due to water draining rapidly from the impermeable surfaces caused by urban development.

The lower sections of north west rivers are wider and deeper with faster currents, carrying a heavy load of sediment, and floodplains which tend to become wider (Figure 7). Where they reach the sea rivers broaden out into estuaries whose shores may be vulnerable to flooding from the sea as well as from rivers. While the middle courses of some streams may have limited areas protected by embankments to keep settlements and

7: The lower section of a north west river; the River Kent. Photograph Ian Whyte.

agricultural land free of flooding, this is much more common and extensive in their lower sections; for example in the Ribble estuary and around Morecambe Bay.

The north west experiences substantially higher average annual rainfall totals than most of the rest of England. Most of the coastal lowlands have at least 700mm of rainfall a year on average and this rises to 3,000mm in the highest parts of the Lake District. The wettest inhabited place in England is Seathwaite near the head of Borrowdale, with an average of 3,556mm a year. Higher up Borrowdale, near Styhead Tarn, it rises to nearly 4,370mm – and even this may be an underestimate.[1] While floods are obviously related to total rainfall they also reflect the rainfall intensity, the speed with which the rainfall occurs. When the intensity is very high, say 55mm or more in an hour, the rain may fall faster than the ground can absorb it causing rapid flooding. What happens to rain once it has fallen depends on the nature of the ground surface. Some of it goes directly into rivers and lakes and is rapidly returned to the sea. Much of it is stored in the soil where it may be absorbed by vegetation and returned to the atmosphere as water vapour. Some of it sinks deeper into permeable bedrock and accumulates as groundwater in underground storage or aquifers. But a good deal percolates through the soil as throughflow and eventually finds its way into streams and rivers. If the soil is saturated, or where the surface is impermeable like a motorway, water will start to flow across the surface of the ground as overland flow.

It is important to remember that the physical extent of even major river floods can be quite small. Within the basin of the River Eden only 143 out of 2,400 sq. km of land are vulnerable to a once in a century flood. In the north west floods can occur at any time of year. Major floods affecting entire river basins are more frequent in winter, though they can occur at other seasons too, while flash floods resulting from localised thunderstorms occur most often in summer and autumn.

The most dramatic changes to rivers and the landscapes around them are caused by very occasional major floods. A flood as severe as the January 2005 Carlisle disaster should only be expected to happen, on average, once every 150 years. This figure is called the return period. It is a measure of probability, not certainty, expressed as the likely frequency of a flood with a particular volume of discharge or greater, or which would cover a particular area, such as a development site. It does not mean that two exceptionally severe floods could not, in theory, occur within a few years of each other – or even in the same season – it is just that this is unlikely.

This assumes, of course, that the probability of a particular level of flood occurring in any year remains constant over time. There are indications that this may not, in fact, be the case. Studies of British rivers have suggested that there were periods in the past which were comparatively flood-rich or flood-poor. The coldest phase of the Little Ice Age in the later seventeenth century may have experienced relatively few major floods. On the other hand the later eighteenth and early nineteenth centuries seem to have witnessed more frequent and more damaging floods.[2] The Tyne certainly experienced an increase in the number and size of floods in the late eighteenth century and between 1840–80 and 1920–50.[3] The frequency of flooding also seems to have increased in upland areas during the late nineteenth century. Such variations could be due to climatic changes but also, possibly, to alterations in land use within river catchments: deciding the relative importance of these influences, in the present or in the past, is a difficult task.

In recent years a specialist branch of research called palaeohydrology has developed which tries to reconstruct how the behaviour of rivers has changed over time through the study of sequences of deposited sediments and other related landforms. Studies of river sediments in upland valleys like the Tyne and Wharfe have identified periods when more material was deposited or eroded as a result of changes caused by human activities like woodland clearance, and by climatic change.[4] Much of this research, however, has focused on the period relatively soon after the last glaciation rather than on recent centuries, although some modern floods have also left significant traces in the sedimentary record.[5] Radiocarbon dating of organic material in river sediments has identified 16 periods of heavy flooding in British river basins in prehistoric times, in several cases associated with phases of large-scale deforestation, as during the Iron Age, The most recent phase, c. 550 years ago, coincides with the start of the Little Ice Age. Historical data can be used to fill the gap between the latest of these flood-rich periods and modern times.

There are links between the occurrence of flooding and the short-term climatic fluctuation known as the North Atlantic Oscillation (NAO), a large-scale contrast in air masses between the sub-tropical high pressure zone and the Icelandic low pressure area. These vary on a timescale from months to several years. The NAO is described as being positive when the Azores High is strong and the Icelandic low is deep, and negative when the reverse situation occurs. When the NAO is positive strong westerly winds develop across the North Atlantic region bringing mild moist air to north west Europe and producing above average rainfall across Britain. When the NAO is negative, characterised by weak pressure gradients and more sluggish westerlies across the North Atlantic it produces colder conditions in northern Europe. Major flash floods in upland catchments tend to be associated with the negative phase of the NAO as this favours the development of localised convectional summer thunderstorms with anticyclonic conditions or by slow-moving near-stationary frontal systems associated with sluggish westerly airflows. This link with the NAO may explain the almost cyclic pattern of flood events in the last 200 years which has been identified in areas like the Yorkshire Dales.[6] A positive NAO with more vigorous westerly airflows is closely associated with winter flooding. Between 1961 and 1990 there was higher average winter rainfall within the north west than in 1941–70 due to increased westerly airflows and a positive NAO.

Floods have been defined in various ways by hydrologists. A frequently-quoted one is 'a body of water which rises to overflow land which is not normally submerged'.[7] Floods may, of course, be measured in other ways than the volume of discharge, such as their financial cost or loss of human life.[8] Flooding might seem to be simply a matter of too much water in the wrong place, but the circumstances which can cause it to be there are remarkably varied. The most common set of conditions occurs with the passage of low pressure systems or depressions in the westerly circulation. These can bring periods of extended rainfall for 24 or even 48 hours across much of the north west, affecting more than one river basin, though often with different levels of intensity. These weather systems tend to be most frequent in the winter half of the year but can occur in summer too. The severity of flooding may be increased when a depression is stalled and becomes stationary, as happened with the floods of June 2007, and also when the ground is already saturated with previous rain so that it cannot absorb any more water.

In the past, especially in the Little Ice Age, many floods were due, in whole or in part, to the melting of snow cover. With the warming of winters in recent decades heavy snowfalls have become rarer in the north west than they were in the eighteenth and nineteenth centuries. Then snowmelt floods seem to have been more common on rivers draining from the Pennines whose broad plateaux provided a more extensive gathering ground for snow than the rugged but less extensive Lake District fells.[9] Nevertheless, the Pennine valley of Dentdale has experienced more than one example of an episode which combined some of the characteristics of a flood with those of an avalanche. This phenomenon, known locally as a gill brak (break), was described in detail by the pioneer geologist Adam Sedgwick who grew up near Dent. In January 1752 a heavy snowfall buried all the streams and gills on the slopes above the Dent valley. Then a sudden rise in temperature caused a rapid thaw as the snow turned to rain. Hard-packed snow in the gills seems to have dammed up the floodwater which finally broke through releasing a mixture of water and ice down on to the farms below. Seven people were swept away and their bodies were later dug out of the snow. Sedgwick remembered a similar, if less deadly, event happening in his youth.[10]

As we have mentioned, localised but often intense convectional rainfall associated with summer thunderstorms, in which the equivalent of a month's rain can fall in an hour or two, can cause severe flash floods, especially in narrow upland valleys. Some devastating ones are described in Chapter 7. When thunderstorms of this kind occur over built-up areas the surface water drains and sewers may be unable to carry off the water fast enough and a short-lived flood can occur in the most unlikely places. One of the most remarkable instances of this was the flooding of parts of the Lancaster University campus, on a hilltop site, following a thunderstorm in October 2000. Runoff from the M6 motorway has also been blamed as a cause of local flooding in villages like Halton.[11] People tend to associate floods primarily with major rivers bursting their banks. In fact, a common cause of floods are small streams and becks which for most of the time seem innocuous and are barely noticed but which, in flood, can cause major problems. Serious floods may also be caused by the inability of surface water drains to carry away rainwater, or the backing up of sewers by flooded rivers. These causes can worsen the problems caused by wider river floods, as happened in Carlisle in 2005. Similar examples in towns like Lancaster are recorded from the late eighteenth century onwards.

Floods can also occur where the urban sections of streams run in culverts if these become blocked or are too small in diameter to cope with the volume of floodwater. The culverted sections of the Dog Beck and Thacka Beck in Penrith have caused flooding in the town since the nineteenth century (Chapter 6), while the Stock Ghyll in Kendal has caused similar problems (Chapter 9).

Much more rarely floods have resulted from the failure of dams. The north west, fortunately, has nothing to compare with the disaster caused by the failure of the Dale Dyke reservoir dam above Sheffield on 11 March 1864 in which 240 people were killed.[12] One major disaster just over the border into Yorkshire occurred on 5 February 1852 when the dam of the Bilberry Reservoir, built only a dozen years before, burst after a week's heavy rain, sending a huge wave of water down the narrow valley towards the village of Holmfirth, drowning over 80 people and demolishing or severely damaging hundreds of cottages, houses, shops, and mills.[13] Nevertheless, a dam burst at Radfield Fold near Darwen in 1848 is claimed to have drowned 12 people.[14] Another dam burst at Church, in East Lancashire, during a flood on 16 November 1866, washing away two cottages whose occupants were only just evacuated in time.[15] The village of Glenridding in Cumbria holds the dubious record of having suffered three floods of this kind caused by the bursting of reservoirs providing water for the nearby Greenside lead mines. The breaching of embankments on the Leeds–Liverpool Canal and the River Douglas in the Rufford area have also caused major flooding in the past.[16]

An unusual and distinctive type of flood is known as a bog burst. This can happen when peat – either level peat in lowland mosses or upland peat on gentle slopes – becomes so saturated with water that it becomes unstable and starts to move. The most famous example is the great bog burst that occurred on Solway Moss in 1771 and which aroused considerable contemporary scientific interest (Figure 8). The event and its after effects were described by the Rev. Dr. John Walker in a paper, accompanied by a sketch map, published in the Royal Society's *Transactions* the following year. Walker appears to have visited the site only a dozen or so days after the first phase of the bog burst. The 'eruption', as contemporaries termed it, was associated with a widespread rainstorm which affected much of northern England on 16–17 November 1771.[17] The flow of liquid peat is said to have covered up to 900 acres of land to a depth of 15 feet or so burying some trees as far as their upper branches. Thirty-five families are thought to have lost their homes as well as their crops and livestock.[18] The Tudor antiquary John Leland, writing around 1533, described a bog burst in 1526 on Chat Moss, between Liverpool and Manchester, near Morley Hall, which, like the event on the Solway, destroyed much agricultural land and killed many fish. The peat was washed into the Mersey and eventually fouled beaches in the estuary over a considerable area.[19] A similar incident is recorded from White Moss near Manchester in 1633 which was described by a contemporary:

> Saturday night this ground brake forth, and, by the violence of the wind and the force of the water which was within, it removed itself; it came in height 4 or 5 yards, and in breadth nearly 20 yards and

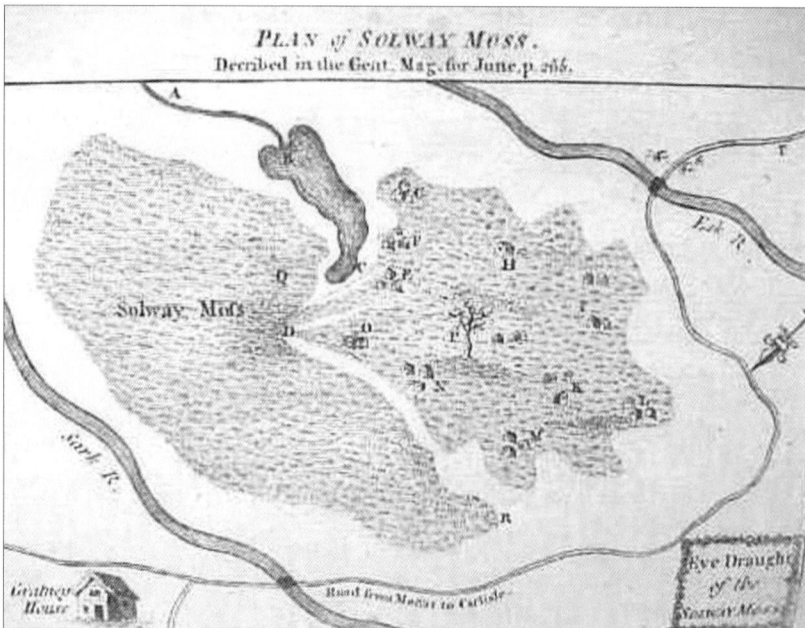

8: Contemporary sketch map of the Solway bog burst, 1771. Photograph Ian Whyte.

sometimes more, and it went violently until it came to a place of descent which we call a clough, and then went down along such place for the space of a mile and a half until it came ... into the River Irk and did raise it as high again as it was before, and so putrified the water that it was as black as a moss pit, and at the Hunts' bank it left I think near a hundred load of moss earth behind it.[20]

The Pennines, with their susceptibility to violent thunderstorms, have produced a number of examples of bog bursts like one near Kirkoswald in 1888. A series of cloudbursts on Knipe Moor near Brough on 18 June 1930 caused a bog burst leaving a series of scars where the peat had been removed down to bedrock and turning the River Eden into a black and filthy mixture which killed thousands of trout as far downstream as Appleby, leaving sticky deposits on the river banks.[21] More recently one occurred in the Pennines in 1983 and was described by an eyewitness as looking like a wall of chocolate sauce sweeping down the valley.[22]

The final cause of floods, an important one in north west England, is flooding by the sea. The coasts of the region, between the cliffs of Great Orme's Head in Wales and St. Bees Head in Cumbia are low and composed of easily-eroded sediments. Flooding along the coasts of Lancashire and around Morecambe Bay has been a feature of human existence from early times. Sand dunes and salt marshes provided some protection for coastal communities but could be breached by high tides and storm surges. The danger of flooding has undoubtedly increased since the eighteenth century with the reclamation of low-lying peatlands in south-west Lancashire, the Fylde and the Lyth valley, as well as the reclamation of land from salt marsh in the Ribble estuary and elsewhere. These newly-won lands were provided with drainage systems which were sometimes expensive and complicated, and were protected by embankments. However, a high tide with strong onshore winds could be a major threat to such lands and the defences were periodically breached (Chapter 8).

Having considered the different types of flooding which can occur in the north west, in the next chapter we look at some of the sources of information which are available about recent floods and those in the more distant past.

Notes

[1]   G. Manley, *Climate and the British Scene* (Collins, 1962).
[2]   M. G. Macklin & B. J. Rumsby, 'Changing climate and extreme floods in the British uplands', *Transactions Institute of British Geographers* (New Series) 32 (2) (2007), p. 175.
[3]   B. T. Rumsby & M. G. Macklin, 'Channel and floodplain response to recent abrupt climatic change. The Tyne basin, northern England', *Earth Surface Processes and*

Landforms 19 (1994), pp. 499–515; M. G. Macklin, D. G. Passmore & B. T. Rumsby, 'Climatic and cultural signals in Holocene alluvial sequences: the Tyne basin, northern England', in S. Needham & M. G. Macklin (eds.), *Alluvial Archaeology in Britain* (Oxbow, 1992), pp. 123–39.

[4]   M. G. Macklin, B. T. Rumsby & J. Heap, 'Flood alleviation and entrenchment. Holocene valley flood development and transformation in the British uplands', *Geological Society of America Bulletin* 104 (1992), pp. 631–43; D. G. Passmore, M. G. Macklin, A. G. Stevenson, C. F. O'Brien & B. A. Davies, 'A Holocene alluvial sequence in the lower Tyne valley, northern Britain. A record of river response to environmental change', *The Holocene* 2 (7) (1992), pp. 138–47.

[5]   Macklin, Passmore & Rumsby, Climatic and cultural signals.

[6]   Macklin & Rumsby, Changing climate and extreme floods, pp. 172–5.

[7]   D. H. Parker, *Floods* (London, 2000) vol 1, pp. 22–4.

[8]   T. Davie, *Fundamentals of Hydrology* (Routledge, 2003), p. 83.

[9]   Archer, *Land of Singing Waters*, p. 6.

[10] Cumbria Record Office (Kendal). Henceforth CRO (K), WDY/193.

[11] G. Maas, *Hydrological and Hydraulic Assessment of Flooding on the River Lune at Halton.* Report Commissioned by the Halton Residents' Group, Lancaster, 2005.

[12] S. Owen (ed.), *Rivers and the British Landscape* (Carnegie, 2005), p. 103.

[13] http://huddersfield/co.uk/huddersfield/holmfirth flood/damage.htm.

[14] L. Markham, *The Lancashire Weather Book* (Countryside Books, 1995), p. 40; W. A. Abram, *History of Blackburn* (Toulmin, 1877), p. 495.

[15] Lancashire Record Office (Henceforth LRO), DP 376/2.

[16] W. G. Hale & A. Coney, *Martin Mere. Lancashire's Lost Lake* (Liverpool University Press, 2005).

[17] Archer, *Land of Singing Waters*.

[18] C. W. J. Withers & L. J. McEwen, 'Historical records and geomorphological events: the Solway Moss 'eruption' of 1771', *Scottish Geographical Magazine* 105 (3) (1989), pp. 149–57.

[19] E. Gorham, 'Some early ideas concerning the nature, origin and development of peat lands', *Journal of Ecology* 41 (2) (1953), pp. 257–74; H. J. Crofton, 'How Chat Moss broke out in 1526', *Transactions of the Lancashire and Cheshire Antiquarian Society* 20 (1902), pp. 139–45.

[20] Crofton, *Chat Moss*.

[21] *British Rainfall*, 1930.

[22] P. A. Carling, 'The Noon Hill flash floods July 17 1983. Hydrological and geomorphic aspects of a major formative event in an upland landscape', *Transactions, Institute of British Geographers*, New Series 11 (1986), pp. 105–18.

CHAPTER 3

Sources of Information on Historic Floods

There is a wide range of information about floods within the north west. Much of this chapter is concerned with evaluating the data which can be obtained from documentary records but we begin by looking at some of the other kinds of evidence that are available.

The sediments deposited by rivers often contain clues to the occurrence and severity of past floods. Major floods may be powerful enough to move what can seem improbably large boulders but once the speed of the current is checked these are dropped rapidly to form spreads which are known as boulder bars. The Mosedale Beck in the Lake District has a spectacular example created by a flash flood in 1749 (Chapter 7). One way of dating flood deposits like these is measuring the size of the lichens on the boulders. Circular lichens grow at a fixed rate which can be calculated for any area by comparing the sizes of lichens on dated surfaces such as tombstones and buildings. These dates can then be calibrated into a graph relating the size of lichens to their ages.[1] In the Lake District this has been done for the Raise Beck, the stream which comes down between Seat Sandal and Dollywaggon Pike to the summit of Dunmail Raise, the pass between Ambleside and Keswick (Figure 9). Measurement of lichens on the boulders on the fan of debris at the head of the pass has indicated five flood episodes, the most recent from the early 1930s and the oldest identified from the 1840s or 1850s, the data fitting in well with known patterns of rainfall.[2]

Turning from the physical evidence of floods to measurements of actual streamflow; for the periods over which they are available, discharge records provide continuous sets of data on river levels. The Environment Agency's website allows you to identify the highest flows for each gauging station.[3] Unfortunately these records often only extend back for 30 years or so. Nevertheless, this covers a number of major (once in a century or longer) floods. For the period before gauged data are available it is possible to identify past floods from entries in compilations of weather data such as British Rainfall (1874–1974). Daily rainfall data are available for the region in continuous series for a number of places from as far back as the later eighteenth century. Particularly wet days or longer periods may provide clues to when floods occurred.[4]

9: Fan of debris on Raise Beck, Dunmail Raise, Lake District. The lichen diameters on the boulders here have allowed some major floods to be approximately dated. Photograph Ian Whyte.

A graphic testimony of the power of past floods are flood marks cut into the stonework of bridges and riverside buildings or preserved on plaques (Figure 10). Individual towns had their own particular yardsticks by which the severity of floods could be assessed. In Carlisle, the height of floods against the Eden Bridge was frequently reported in local newspapers. In Kendal the levels of floods were recorded in relation to how far up Lowther Street the water reached, measured by the number of railings reached by the water (Figure 11). Care is needed in interpreting flood marks though because if the bed of a river has changed over time (by having gravel dredged from it for example) then the volume of water needed to reach a particular level will not remain constant.[5]

If you want to go back beyond these sources, or wish to flesh out the picture of the scale and impact of recent floods, a range of documentary sources can be used.[6] We saw in Chapter 2 how floods are defined by hydrologists. Their definitions are not necessarily the same as those used in historical sources. A working definition for the latter might be 'an overflow of water from normal river channels which, by causing sufficient damage and inconvenience, attracts enough attention to be recorded in a historical source'. This is a more practical but at the same time more variable definition. For example, a major rise of river levels beyond normal channels which did not attract attention by causing sufficient damage might not be recorded in any historical sources. A flood was not automatically a hazard under such circumstances. Given that the volume and range of historical sources falls off the further you go back in time, this helps to explain why the incidence of floods recorded in documents diminishes rapidly for periods before the eighteenth century. While there may indeed have been fewer major floods in the

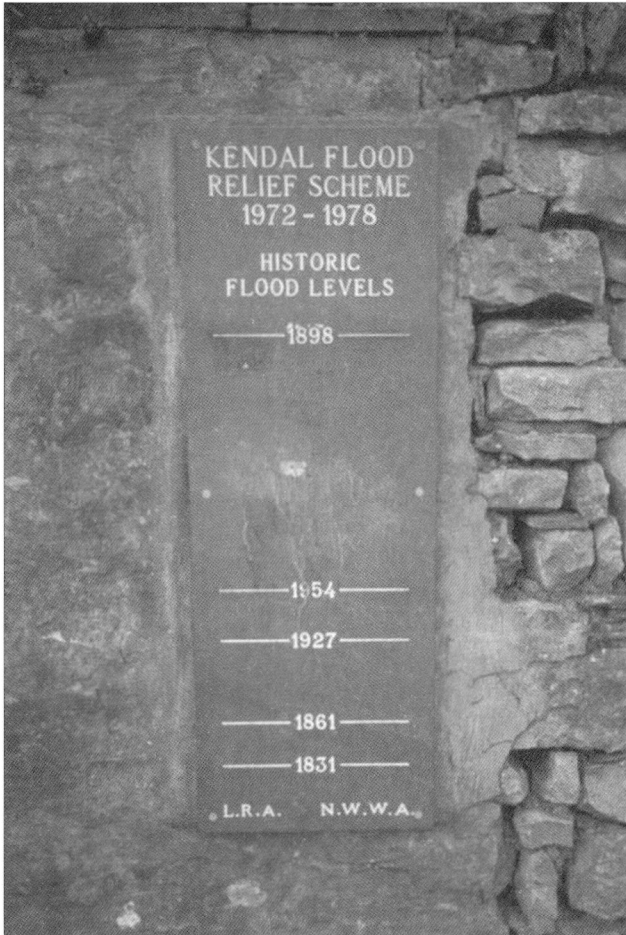

10: Plaque marking heights of some severe floods in Kendal. Photograph Ian Whyte.

seventeenth century than in the eighteenth due to lower rainfall levels it is also likely that the number of floods recorded was smaller because the ones which did occur caused less damage. This was due in turn to the more limited amount of development on flood plains at this time. In the narrow upland valleys of east Lancashire there were many floods but relatively little flood hazard before the Industrial Revolution led to the building of textile mills and associated workers' housing on the valley floors. In the seventeenth and eighteenth centuries a flood which merely inundated areas of wet riverside meadow might not have been seen as worth recording in most sources and might only appear, almost randomly, in a letter or diary entry. In the nineteenth century, as flood plains were increasingly developed for industry, housing and communications, levels of flooding which would have caused few problems in previous centuries started to become more damaging and more newsworthy.

Of the wide range of historical sources relating to floods that is available newspapers are perhaps the most useful and certainly the most voluminous. They have been used in a number of previous studies of flooding in Belfast,[7] Cork[8] and the Swansea valley.[9] An online index for *The Times* is available extending from 1785 until 1985.[10] However, national newspapers only chronicle the most severe events and usually not in very great detail. Local newspapers are much more informative and are available on microfilm, and sometimes in their original form, in larger reference and local studies libraries. Unfortunately, they are not often indexed although some reference libraries keep files of press cuttings relating to floods. If you have a date for a particular flood from another source it is easy to check contemporary newspaper reports but to scan your way through a few years or even months of a local newspaper in search of floods, particularly as most of them are only available on microfilm, is a tedious and eye-straining process.

11: Lowther Street, Kendal. Local people measured the severity of floods by how far up the street the water reached. Photograph Ian Whyte.

Newspaper reports of floods often compared them to earlier disasters which can be followed up in turn. In early times, however, there could be difficulties due to variations in the quality of reporting. On 29 September 1809 *The Times* reported a destructive flood in the lower Eden valley. The paper was unable to comment on the situation in the nearby Tyne basin, other than to report a rumour that the bridge at Hexham had been washed away because their Newcastle correspondent had not bothered to report any news of flooding. Clearly there was a hit or miss element to the recording of such events. There is also a problem with identifying severe thunderstorms and resulting flash floods. Because of their localised nature they were not always mentioned even in local newspapers.

Early newspaper reports of floods tend to confine themselves to mentioning damage to bridges and the flooding of towns.[11] The location of a flood might determine whether or not it was recorded in particular newspapers. The severe flash flood of 1889 in the Pennine valley of Garsdale (Chapter 7) received scant attention in the *Westmorland Gazette* although only a few miles from Kendal, probably because it was located in Yorkshire. The bursting of the Keppel Cove dam in 1927 and the resulting flooding of the village of Glenridding with serious (if highly localised) damage, did not make much impact on the national press, partly because it coincided with a much more widespread flood at Fleetwood (Chapter 8). The event, reported in a small paragraph in *The Times*, was described as 'Lake empties' rather than 'village devastated'. This highlights a perennial drawback of newspaper reports; events were screened and filtered by editors who were often based outside the area in which the floods occurred and whose views and interests are likely to have been different from those of the people who actually experienced

12: Floods in Salford in 1946.

the event. Ultimately the best source for individual and community responses are diaries and autobiographies or, in more recent times, interviews.

Old photographs from the later nineteenth century onwards can provide vivid images of the levels reached by floods and the responses of the people affected (Figure 12). Many photographs reproduced in newspapers are of poor quality but local history and reference libraries may have the originals, though it is important to remember that they were not necessarily always taken when floods were at their peak levels.

For the pre-newspaper era periodicals contain references to unusual environmental occurrences such as earthquakes, storms and floods. One of these is the *Gentleman's Magazine*, starting in 1731, which can be searched online.[12] Among other entries it has a graphic account of the flash flood that devastated the Vale of St. John in the Lake District in 1749 (Chapter 7). More modern magazines like *Lancashire Life*, *Cumbria* magazine and *The Dalesman* also contain occasional articles on past floods.

For the period before newspapers, and as supplementary evidence even when they are available, other historical sources can provide valuable information. The county Quarter Sessions were responsible for maintaining the more important bridges and contain information on the damage or destruction caused by floods, often picking out seventeenth or eighteenth century floods not mentioned in other sources. How big a flood was needed to demolish a particular bridge would have depended partly on its age, size, quality of design and state of repair as well as the severity of the flood. Bridges varied in size and strength from foot and farm bridges carrying purely local traffic through small stone packhorse bridges to more substantial structures which were maintained by the county and were proof against all but the most severe inundations. The regimes of north west rivers with their sudden floods and rapid peaks, may help to explain why there are so few medieval stone bridges left, though the poverty of the region in the past is also likely to have been an influence. In the old county of Westmorland, for example, there are only two pre-seventeenth century bridges, at Kirkby Lonsdale and Warcop (Figure 13).

Parish registers and other local records such as churchwardens' accounts may also contain information on floods. Parish burial registers rarely listed causes of death but they do sometimes mention when people died in unusual circumstances. It is likely though that far more people

13: The late medieval bridge at Warcop is the oldest on the River Eden. Photograph Ian Whyte

drowned in rivers and lakes through boating accidents or falling in and being unable to swim than were washed away in floods. Nevertheless, parish registers occasionally contain marginal notes and entries referring to notable floods like the register for Hawkshead which records for 10 June 1686 a fearful thunderstorm and flood the like of which could not be remembered by anyone alive. It washed away houses, mills and several bridges, depositing great beds of sand and spreads of boulders, and carrying away sections of road.[13] Churchwardens' accounts sometimes give information on payments to flood victims.

Guidebooks and topographic descriptions mention floods, sometimes in graphic detail. They tend to focus particularly on events resulting from severe localised thunderstorms or 'waterspouts' as they were called in the eighteenth century. William Gilpin, for example, gives a detailed description of the storm over Grasmoor which flooded the Vale of Lorton in 1760. At the time of his visit, 12 years later, the damage caused by the flood was still clearly visible. Many guidebook entries of this kind are, however, derivative and second hand. Eighteenth-century British printed books can be accessed online.[14]

Among private records estate papers may contain information relating to repairs resulting from flood damage. As early as 1296, for the de Lacy estates between Clitheroe and Rossendale, the accounts refer to the repair of flood damage by the River Calder.[15] Correspondence and diaries also mention floods but the quality of detail varies. In the diary kept by Tom Rumney of Mellfell in 1805 and 1806 there is not enough detail to indicate whether he is simply describing an unusually wet day or a genuine flood.[16] On the other hand the diary of Isaac Fletcher of Underwood, near Cockermouth, does provide sufficient detail to make it clear when he is talking about a real flood.[17] The diary of Timothy Cragg of Ortner in Wyresdale records major floods, often in some detail, for the Wyre and Lune valleys from the 1780s to the early nineteenth century.[18]

The best sources are often the unexpected 'gems' such as a letter from a lady in Grasmere providing a first-hand account of the flood of 1898.[19] Even better are the records of flood relief organisations. The Garsdale Inundation Fund, set up by local people after a flood in 1889, detailed the damage and the estimated costs of repair for each farm in the valley.[20] The relief fund established in Walton le Dale after the severe flood of 1946, and at Oswaldtwistle following flood damage in 1964, required each claimant to fill in a form detailing the items destroyed and damaged, providing a wealth of detailed information.[21] The level of detail supplied, and the personal comments made on some of the forms bring us closer to the concerns and fears of the flood victims than any other source. No comparable set of data has yet been found for the 1927 sea flood in Fleetwood but a minute book of the flood relief committee shows, in broad terms, how the money that was donated was allocated.[22]

Local reference libraries often keep files of press cuttings and other information on past floods. Fleetwood has folders of cuttings relating to the disastrous flood of 1927. A useful file of material relating to floods in Kendal is reserved in the Cumbria Record Office there. It was produced by Paul (later Lord) Wilson, councillor and mayor of Kendal in the period from the 1950s to the 1970s.[23] As an engineer he took a keen interest in the meteorological background to flooding in Kendal and to the features of the catchment and channel of the River Kent which he considered had exacerbated the problem. The file contains his own written records, based on personal observation and a search through newspapers, relating to levels of rainfall and the height of various floods going back into the late eighteenth century.

Lastly, maps are a major source of information on changes in land use on river floodplains, as well as on variations in the course of streams resulting from major floods. A key source is the first edition of the six-inch to the mile Ordnance Survey map, dating from the late 1840s in Lancashire and the early 1860s in Cumbria and available online,[24] together with subsequent editions from the later nineteenth and early twentieth centuries for comparison. Other accurate large-scale maps which may be of use are the tithe surveys of the 1830s or maps of areas enclosed under parliamentary act in the late eighteenth and early nineteenth centuries which are kept in county record offices.

How did the occurrence of floods in north west England change over the last few centuries? Figure 14 shows the chronologies of KNOWN major and moderate river floods obtained from historical records and other data, for the region. Minor floods have been omitted as these are much more likely to be recorded in modern times than in the past,

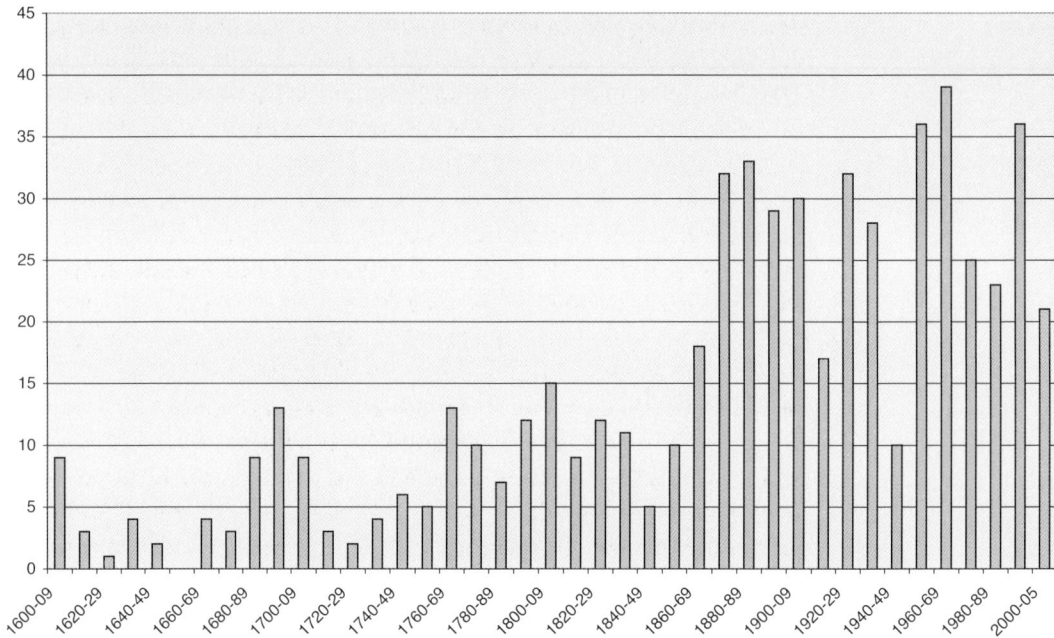

14: Major and moderate floods in the north west AD 1600–2005. Ian Whyte.

particularly the more distant past where the historical record becomes much thinner, so distorting the picture unduly. The pattern of sea floods will be discussed in Chapter 8.

The most obvious trend in the graph is a marked increase in the total number of recorded floods over time. As mentioned above, this undoubtedly reflects in part improvements in the volume and quality of source material.[25] It is probable that due to limitations in the quality and quantity of the data the graphs do not fully reflect the occurrence of floods before the eighteenth century. Few floods are recorded at all before 1600. However, the upward trend of the graphs over time is also likely to reflect the increasing development of housing, factories, roads and improved agriculture on river floodplains from the later eighteenth and particularly the mid-nineteenth century onwards. This trend is especially evident in urban areas but also affected the countryside. While the floodplains of rivers like the Irwell, flowing through the heart of Manchester, were increasingly built up there was also a rise at the same period in the number of rural factories in areas like south Cumbria making textiles, bobbins and gunpowder, as well as much reclamation around Morecambe Bay, in the Fylde and in south west Lancashire, land which was vulnerable to flooding and protected by drainage systems and embankments.

It is likely too that changing public perceptions and lifestyles have affected the reporting of floods. The flooding of an early nineteenth century cottage containing few possessions and mainly wooden furniture would have caused little lasting damage and been less newsworthy than the flooding of its modern equivalent with fitted carpets, soft furnishings, electrical wiring and consumer goods.[26] These variations may also reflect climate changes with significant fluctuations in the amount of rainfall over time, particularly with the shift from a colder seventeenth century to a wetter later eighteenth to mid-nineteenth century. When the graph is examined on a timescale of decades rather than centuries it is clear that there are marked short-term variations in the incidence of flooding which cannot be explained by differences in the source material. There are what have been called flood-rich and flood-poor periods. Flood-rich periods occurred in the mid/later seventeenth, early eighteenth, late eighteenth, early nineteenth and late nineteenth centuries. In the twentieth century there was a peak from the 1920s to the 1960s. Flood-poor periods included the mid-eighteenth and mid-nineteenth centuries, the 1910s and the 1970s. In particular, despite possible perceptions to the contrary, the period from the 1970s to the present day appears to have been less flood-prone than earlier decades. There is a lot of similarity between flood-rich periods in the north west and ones which have been recorded for rivers in the north east from Northumberland to Yorkshire.[27]

Which was the worst recorded flood in the north west? This is a difficult question as the flood with the greatest discharge was not necessarily the most damaging. Nor was the date of the worst known flood the same in each catchment. The worst river flood from the Ribble to the Irwell and Mersey was probably that of November 1866. On the Eden, however, January 2005 was the worst, followed by March 1968 and February 1822. On the River Kent, by contrast, November 1898 was probably the worst flood on record, followed by August 1890, while on the Lune the worst two were September 1892 and January 1995. In terms of sea floods the one of 1720 was almost certainly the most extensive and damaging.

In this chapter we have reviewed the historical sources for flooding in the north west and the patterns over time which can be discerned from them. Now we turn to the impacts that these floods had on communities and how these too varied over time.

Notes

[1] A. D. M. Harvey, R. W. Alexander & P. A. James, 'Lichens, soil development and the age of Holocene valley floor landforms, Howgill Fells, Cumbria', *Geografiska Annaler* 66A (1984), pp. 353–66.

[2] R. M. Johnson & J. Warburton, 'Flooding and geomorphic impacts in a mountain torrent: Raise Beck, Central Lake District', *Earth Surface Processes and Landforms* 27 (2002), pp. 945–69; M. G. Macklin & B. T. Rumsby, 'Changing climate and extreme floods in the British uplands', *Transactions Institute of British Geographers* (New Series) 32 (2007), pp. 168–86.

[3] www.environment-agency.gov.uk/hiflowsuk/

[4] P. A. Barker, R. L. Wilby & J. Borrows, 'A 200–year precipitation index for the central English Lake District', *Hydrological Sciences Journal* 49 (2004), pp. 769–85.

[5]   D. R. Archer, F. Leesch & K. Harwood, 'Assessment of severity of the extreme River Tyne flood in January 2005 using gauged and historical information', *Hydrological Sciences Journal* 52 (5) (2007), pp. 992–1003.

[6]   L. J. McEwen, 'Sources for the establishment of historic flood chronologies (pre-1970) within Scottish river catchments', *Scottish Geographical Magazine* 103 (1987), pp. 132–40; H. R. Potter, 'The use of historic records for the augmentation of hydrological data', *Institute of Hydrology Report* no.46 (1978); A. C. Bayliss & D. W. Reed, 'The use of historical data in flood frequency estimation', *Centre for Ecology and Hydrology* (2001).

[7]   D. B. Prior, & N. L. Betts, 'Flooding in Belfast', *Irish Geography* 7 (1974), pp. 1–18.

[8]   J. G. Tyrrel & K. J. Hickey, 'A flood chronology for Cork city and its climatological background', *Irish Geography* 24 (1991), pp. 81–90.

[9]   R. P. D. Walsh, R. N. Hudson & K. A. Howells, 'Changes in the magnitude and frequency of flooding and heavy rainfall in the Swansea valley since 1875', *Cambria*, 9, (1983), pp. 36–60.

[10]   http://infotrac.gategroup.com/menu

[11]   Archer, *Land of Singing Waters*, p. 92.

[12]   http://www.bodley.ox.ac.uk/ilej/journals/srchgm.htm

[13]   H. S. Cowper, *The oldest register of the parish of Hawkshead in Lancashire 1568–1704* (Bemrose, 1897), pp. lx–lxi.

[14]   http://galenet.galegroup.com/servlet/ECCO?locID

[15]   M. A. Atkin, 'Land use and management in the upland demesne of the De Lacy estate of Blackburnshire *c.* 1300', *Agricultural History Review* 42 (1) (1994), pp. 1–19.

[16]   A. W. Rumney (ed.), *Tom Rumney of Mellfell (1764–1835) by Himself as set out in his Letters and Diary* (Titus Wilson, 1936).

[17]   A. J. L. Winchester (ed.), *The Diary of Isaac Fletcher of Underwood, Cumberland, 1756–1781* (Cumberland and Westmorland Antiquarian and Archaeological Society Extra Series, 1994).

[18]   LRO, DDX 760/1.

[19]   CRO (K), WDX 393.

[20]   CRO (K), WPR 60.

[21]   LRO, UDWd 69, UDOs 18/16.

[22]   LRO, MBF/11/11.

[23]   CRO (K), WD/PW.

[24]   www.old-maps.co.uk

[25]   R. Brazdil, G. Rudiger, C. Pfister, P. Dobrovolny, J-M, Antoine, M. Barriendos, D. Camuffo, M. Deutsch, E. Guidobone, O. Kotyza & F. Rodrigo, 'Flood events of selected European rivers in the sixteenth century', *Climatic Change*, 43 (1999), pp. 239–85; N. Macdonald, A. Werrity, A. R. Black, & L. J. McEwen, 'Historical and pooled flood frequency analysis for the River Tay at Perth, Scotland', *Area*, 38 (1) (2006), pp. 34–46.

[26]   E. C. Penning-Rowsell & J. W. Handmer, 'Flood hazard management in Britain: a changing scene', *Geographical Journal*, 152 (2) (1988), pp. 209–20.

[27]   Archer, *Land of Singing Waters*; B. T. Rumsby & M. G. Macklin, 'Channel and floodplain response to recent abrupt climatic change. The Tyne basin, northern England', *Earth Surface Processes and Landforms* 9 (1994), pp. 499–515.

CHAPTER 4

The Changing Impact of Flooding on Communities in the North West

The impacts of flooding are wide-ranging and in the north west have varied considerably over time as the economy, standards of living and people's attitudes have changed. The effects of floods also depend on how people react to them as a potential hazard. Studies of people's perceptions of the risks of environmental hazards have shown that they tend to interpret the information available to them very selectively and usually substantially underestimate the level of risk. People's perceptions of, and responses to, flood hazards have changed through the centuries too. Farming communities in the past were better prepared for floods and more resilient in the face of their effects than modern urban dwellers. Attitudes to flooding and demands for flood protection have been influenced by religious beliefs, levels of material wealth and perceptions of authority. The increase in the range and cost of people's household possessions, including fitted carpets and electrical goods, has also influenced perceptions. In the past floods were viewed as acts of God and even punishments for sin. The development of science from the eighteenth century altered people's perceptions and floods were increasingly seen as Acts of Nature. Only from the later eighteenth century, parallel with the Enlightenment and the rise of modern science with the idea that human society can control nature, did the first large-scale flood protection schemes develop in Britain. These were extended in the nineteenth and twentieth centuries (Chapter 10). The occurrence of flooding tends to be seen today as the result of the failure of flood defence systems and of those responsible for installing and maintaining them. More recently still floods have been viewed as the result of human activity. People's reaction to disaster has shifted from one where no one is to blame to the modern view that someone is responsible, and possibly criminally liable, for any disaster which occurs.[1]

Floods are usually portrayed as disasters, and for most people they undoubtedly are and were, but it should be remembered that they produce gainers as well as losers. Following the January 2005 flood in

Carlisle owners of rented property, builders and tradesmen, car sales firms and shops supplying electrical goods and furniture all profited from the damage caused. In a similar way concern over the damage caused by the flood of 1898 in Grasmere was balanced, among labouring families, by the knowledge that repairing the damage caused to the main roads would provide work for them through the winter.[2]

Physical effects of flooding

One of the most dramatic results of serious floods is their physical impact on the landscape. The changes caused by floods in steep-sided upland valleys can be dramatic. The Howgill Fells, between the Lake District and the Yorkshire Dales, are a good example. In valleys like Carlingill the headwaters of streams are scarred by systems of gullies, Many are grassed over, the product of ancient phases of flooding and erosion, and are not active under modern conditions. Some of them may relate to the effects of the clearance of woodland cover in medieval or even prehistoric times. Other gullies and erosion scars are bare and still active. A flood like the one which hit parts of this area in 1982 can reactivate these old gully systems. The material which was washed out of them was deposited lower down the slope as alluvial fans or in the main valleys.[3] In 1898 so much sediment was washed down from Dunmail Raise by flooding that Grasmere was turned red with it.[4] Floods can also erode and re-deposit waste materials from former mining activity causing heavy metal pollution to riverside pastures, as occurred in the Vale of York following the severe floods in 2000.

We have already seen that floodwater can carry large boulders along and drop them in spreads or boulder bars. A combination of heavy rain, saturated ground and a steep slope can lead to landslides during flood conditions. The A591 road along Thirlmere below the western slopes of Helvellyn is particularly prone to being blocked in this way. A walk along the footpath which runs parallel to the road at a higher level shows why; the slope is dissected by a number of steep gills, many of which are dry for part of the year. During floods, however, they bring down huge spreads of loose boulders, stones and gravel which are unstable on the steep slopes, ready to slide when lubricated and often block the road, as happened in 1995, 1997 and 2000 (Figure 15). Railway embankments are also prone to landslides when saturated and there have been two derailments on the Settle–Carlisle line as a result of this in recent years, in 1995 and 1999.

Floods can sometimes make streams change their courses. One of the most dramatic examples is at Legburthwaite in the Vale of St. John where the flash flood of 1749 diverted the stream into an entirely new rock-cut channel. A thunderstorm over the headwaters of the River Gelt in north

15: Steep, loose flood debris on slopes above Thirlmere, ready to slump down and block the main road. Photograph Ian Whyte.

Cumbria in August 1894 produced a flash flood which caused the river to take a completely new course just before it joined the River Irthing. Another example is Raise Beck in the Lake District (see Chapter 3). At the top of Dunmail Raise the stream has deposited a fan of debris. Major floods in the past have changed the stream's direction; it has sometimes flowed northwards towards Thirlmere but at other times south to Grasmere. Before the building of the Thirlmere dam in the 1890s Raise Beck flowed down to Grasmere. The flood of 1898 diverted it north to Thirlmere but in the early twentieth century other floods shifted it back south again then north until 1995 when a flood sent it towards Grasmere once more. After this the stream channel was altered by engineering works so that it should flow northwards permanently. Its abandoned channel, on the south side of the watershed, can still be clearly seen (Figure 16).[5]

Mortality

Perhaps surprisingly, floods do not appear to have caused many fatalities in the past. This may have been linked to the care with which sites of farmsteads and cottages were chosen. The two worst single incidents identified relating to river floods were one on the Eden in 1687 which struck the village of Arthuret in the middle of the night drowning two women and five children[6] and the curious flood/avalanche which occurred in Dentdale in January 1752 in which seven people were killed when their house was overwhelmed.[7] The worst loss of life on record was probably the sea flood of 18–19 December 1720 when 20 or more people are thought to have drowned along the Lancashire coast.[8]

16: The abandoned southward-flowing channel of the Raise Beck, Dunmail Raise. Photograph Ian Whyte.

Apart from unusual episodes like this the greatest danger during floods was probably to farmers trying to rescue livestock or to travellers being swept away attempting to cross swollen streams. The great flood of 1771 resulted in three men drowning at Wennington when a bridge collapsed under them. Loss of life by drowning was probably more common among travellers crossing Morecambe Bay. At least 100 people are thought to have been drowned since the sixteenth century making this frequently-used but dangerous crossing.[9] The worst single drowning incident on record in the region was in 1635 when the Windermere ferry sank with a loss of 47 lives. Fatalities could also arise indirectly from flooding as with the walker who was killed by a fall in Langdale in 1966 after a flood had washed out the footpath.[10] Deaths might also arise from damage due to the gales which sometimes accompanied heavy rain; in 1893 three girls were killed at Cowan Head Mill when a brick chimney collapsed in a storm.[11]

This amounts to some 125 known fatalities over four centuries or so. The tendency has been for mortality to fall since the nineteenth century but the loss of life in the floods of summer 2007 shows that this trend may not automatically continue. A number of near escapes are also recorded including a shepherd nearly drowned at Keswick in 1794[12] or two people swept away at Marron Bridge in west Cumberland in 1711 who managed to cling on to some bushes further downstream.[13] One might also include indirect deaths resulting from flooding such as Elizabeth Benson who was drowned in 1762 crossing Birker Beck, Eskdale at the site of a footbridge which had been destroyed by a recent flood.[14] To judge by the entries in parish registers the numbers of people drowned in floods was greatly exceeded by those lost in lakes and rivers

17: The old road up Great Langdale, well above reach of valley-floor flooding. Photograph Ian Whyte.

due to bathing, boating and fording accidents under normal water levels. In the registers of Hawkshead parish for the late sixteenth and seventeenth centuries only one recorded drowning out of 14 was (possibly) flood-related.[15]

The events of January 2005 show that floods still represent a danger to life even though two of the fatalities in Carlisle were due to heart attacks rather than drowning. The *Environment Agency Catchment Flood Management Plan for the River Eden*[16] suggests that the flood risk to people in the future is likely to be greater in smaller, steeper stream catchments rather than major rivers but only about 10 per cent of the known fatalities fall into this category. Several major severe flash floods in upland catchments like those of 1749, 1760, 1927 and 1967 were not associated with any deaths.

Agriculture and rural areas

Today 90 per cent of all properties affected by flooding are in urban areas.[17] This figure was much lower in the past due to the smaller size of towns and less flood plain development in urban areas. In some respects the inhabitants of the countryside were better able to cope with flooding in the seventeenth or early eighteenth centuries than in modern times. Roads and farmsteads were often located on the valley sides well above the reach of floods from the main streams. The old road up Great Langdale from Chapel Stile and the one from Rydal to Grasmere, the main road during the period of Wordsworth's stay at Dove Cottage, are good examples (Figure 17).[18] Both roads and steadings may have been

Table 1. Fatalities in North West Floods.

1650	Walton le Dale	Woman drowned in flooded mill race
1697 Nov 11	Arthuret, Carlisle	2 women, 5 children
1720 Dec 18–19	N. Meols, Fylde, Cockerham	20+ drowned – sea flood
1737 Dec	Preston	Robert Heath swept away while leading horse through floods near Ribble Bridge
1752 Jan	Dentdale	7 people – 2 families – died
1768 Aug	Leyland	One family in house
1771 Nov 17	Wennington	3 men drowned when bridge collapsed
1777 Jul 29	Holmfirth	3 men drowned in flood
1783 Aug	River Lostock	George Charnel drowned in flood
1787 end of June	Salford	Salford Bridge washed down – man on it drowned.
1798 Autumn	Irwell	Coach and horses swept away – 'men and women drowned'
1806	Netherhall, W Cumberland	Boy drowned in R. Ellen
1809 Sep 26	Maryport	Boy and woman drowned when bridge collapsed
1814 Dec *c.* 20	Bampton	1 poor woman drowned
1817 Jan 4	Eden, Nunnery	1 exciseman; Mr Robson, Kirkoswald, drowned
1821 Nov – end	Between Keswick and Bassenthwaite	1 clergyman washed off horse and drowned.
1829 Oct 14	Near Brough, Westmorland	Carriers' wagon from Stockton to Penrith overturned – 2 people drowned
1831 Nov 25	Burrow Beck	Elizabeth Postlethwaite drowned
1837 Dec 21	Mersey	Baby in cradle observed being washed downstream
1840	Holcombe Brook	Carter, woman and child drowned
1848	Darwen	Dam burst – 12 people drowned
1861 Dec 25	Lowther Br	Rob Parker drowned
1866 Nov 16	Manchester, Little Broughton	4 people drowned
1868 Feb 1	Keswick	Mr Hayton, tradesman, drunk – R. Greta
1870 Jul 9	Dentdale	2 people drowned in flash flood – one in railway tunnel the other trying to cross beck
1872 Jul 13	Manchester	1 person drowned
1872 Jul 26	Caton	1 child drowned
1872 Jan 21	Keswick	1 man drowned in R. Greta
1874 Oct 6	Cockermouth	Eliz Steward, hawkers's wife, drowned washing clothes
1881 July 6	Bacup	Man, granddaughter and 2 women drowned
1882 Nov 1	Cleator	Boy drowned
1892 Sep 2	R. Petteril, Carlisle area	2 boys drowned
1926 July 19	Gimmer Crag, Langdale	Climber washed off crag in torrential rain
1926 Sep 21	Wigton	Boy drowned
1927 Oct 29	Lancaster	3 people drowned
1927 Nov 1	Fleetwood	6, including mother and 3 children, drowned in sea flood
1927 Nov 3	Morecambe	1 man killed
1933 Jul	Liverpool	Cyclist drowned
1942 Sep 5	Halton railway station	Thomas Lamb, age 9, drowned
1944 May 29	Holmfirth	3 men drowned

1953 Aug 3	Troutbeck	1 man drowned
1954 Jun 16	Killington	1 child drowned
1964 July 18	Haslingden	1 woman drowned
1964 Dec 12	Austwick	1 man drowned
1965 Nov 1	Shap	1 person drowned
1966 Aug 13	Langdale	1 person killed as result of path being swept away by flood
1968 Mar 24	Dufton Pennines	2 walkers on Pennine Way died
1995 Feb 1	Aisgill	1 killed in train derailment caused by flood-related landslide
1998 Aug 2	Langwathby	Woman died of heart attack
2005 Jan 8–9	Carlisle	3 people died

vulnerable to localised flooding by tributary streams and gills under exceptional conditions, as happened in Great Langdale in 1966, but this would have been exceptional. One of the distinctive features of the Wray flood of 1967 was that it damaged seventeenth-century farmsteads and cottages which had never previously been flooded.[19] Wordsworth, in his *Guide to the Lakes* of 1810, commented on the locations chosen for traditional Lake District farmhouses; in sheltered positions on slopes, but above the valley floods.[20] Some of the villas and mansions built during the nineteenth and early twentieth centuries by incomers may have been less well sited from the point of view of flood protection.

Wet riverside ground was used as natural hay meadow in the past while better-drained floodplain land would also have been used for growing cereals. The main flooding period, during the winter, occurred after the harvest and before the sowing of oats and barley, the main crops except in south Lancashire. Summer floods were potentially bad because they might flatten or wash out crops or even bury them with sand and gravel. On the other hand winter floods in the south of the region might wash out winter-sown wheat or root crops like potatoes and turnips that were already in the ground. The return of so much land to permanent pasture in areas like the Lyth valley since the Second World War, and the shift from hay to silage production, has probably greatly reduced potential flood damage to crops in recent decades. Down to the mid seventeenth century there was a real danger that a crop-damaging flood might actually bring food shortage and starvation.[21] Hay meadows were equally vulnerable to summer floods though really serious winter ones might wash entire haystacks away. Hay which was harvested wet would have been less nutritious and more liable to damage by mould. The areas where crops were vulnerable to flooding were on lowland floodplains, as around Carlisle. Wet valley bottom washlands, as on the Kent between Staveley and Kendal, soaked up floodwater, delaying the onset of floods and reducing their severity.

18: View from old road over marshy floor of Great Langdale. Photograph Ian Whyte.

Important as crops were, much of the north west was pastoral country and livestock figure prominently in most reports of major floods before the mid-twentieth century, with accounts of sheep, cattle and even horses being swept down flooded rivers and of heroic efforts to rescue threatened animals. It is easy to forget how many animals were kept in urban areas into the later nineteenth century; chickens, ducks and especially pigs were widely kept. The *Preston Guardian*'s account of the flood of November 1866 tells how householders near the Ribble Bridge were forced to save their pigs by getting them out of their sties and taking them into their upstairs bedrooms until the water receded.[22] When Walton le Dale was flooded in 1946 some of the people who claimed for damages clearly kept poultry on a significant scale to offset the effects of egg rationing.[23]

The housing of cattle in many parts of the region into the eighteenth century would also have kept some of the livestock away from the most flood-prone locations at the most dangerous time of the year.[24] In winter, however, some cattle and sheep were grazed on low-lying riverside pastures and, unless removed in time, were vulnerable to being washed away and drowned. Many reports of nineteenth-century floods refer to the sight of animals, dead or alive, being swept downstream.

Floods also caused considerable damage to walls, hedges, fencing and local bridges.

As we move from the nineteenth to the twentieth century there are some marked differences in the kinds of damage caused by floods as a result of changing technology and lifestyles. Reports of floods in the early twentieth century start to mention telephones and electricity supplies being affected by flood water. Railway embankments and cuttings were especially vulnerable to landslides, blocking the line in cuttings or leaving the tracks hanging in mid air when embankments collapsed. The period after the First and especially the Second World War also saw the start of reports of flood damage to caravan parks and especially campsites. It seems to have been assumed that, because caravans or tents were not permanent structures, they were less flood-prone than houses and could, as a result, be sited close to major streams without any qualms of conscience. There are even hints in the reportage that being flooded out of a tent, forced to evacuate a campsite in a matter of minutes, and have all one's possessions soaked or even washed away by a flood, was all part of the 'fun' of camping. The rising ownership of motor vehicles led to increasing problems of their being immobilised by floodwater which would not necessarily have caused problems to a horse and cart.

As well as damaging agricultural land, floods were also capable of washing away entire cottages and houses. This was probably less frequent in upland areas where houses were built of stone than in the lowlands where timber-framed buildings and even ones with walls of solid clay were still found into the nineteenth century. This may explain why the sea floods of 1720 were so damaging, supposedly washing away 157 houses and damaging 200 more.[25] Even more robust structures like churches were not immune to serious damage; in 1744 the church at Garstang had to be rebuilt after flood damage.[26] As housing standards improved such instances became rarer but as late as 1891 in Darwen a number of houses in the Grimshaw Street and Pilkington Street area collapsed and were washed away by a flood.

Communications and Bridges

Perhaps the most common flood impact was the damaging or destruction of bridges. This was particularly likely to occur if a bridge became submerged to, or beyond, the top of its arches and especially if debris was battered against it by the floodwater (Figure 19). It is no coincidence that surviving old bridges, like the Devil's Bridge at Kirkby Lonsdale tend to be ones with very high arches (Figure 20).

Many small local bridges providing access to farms and hamlets were washed away by floods without record. The disasters to bridges which were recorded in the Quarter Sessions were those which carried

19: Detail of Eamont Bridge near Penrith showing rebuilding, possibly after flood damage. Photograph Ian Whyte.

20: The late medieval high-arched 'Devil's Bridge' at Kirkby Lonsdale, still standing despite numerous floods. Photograph Ian Whyte.

significant volumes of through traffic. Petitions to the Westmorland Quarter Sessions in 1750 from local farmers and landowners in Longsleddale relate that the bridges at Sadgill and Wadshaw had been destroyed by floods the previous August. Wadshaw Bridge was described as being on the high road from Ambleside to Appleby. The petitioners claimed to have built the bridge 24 years before at their own expense but now wanted county money to reconstruct it.[27] It was common for bridge maintenance to be skimped before the nineteenth century and this is likely to have hastened the destruction of many of them.

Some periods come out as having particularly large numbers of bridges destroyed, such as the mid-late eighteenth century and especially the

21: Site of bridge in Great Langdale demolished by the January 2005 flood. Photograph Ian Whyte

early nineteenth century while other periods have little recorded damage, like the mid seventeenth, mid nineteenth and early twentieth century. It is clear that the number of bridge failures has fallen in recent times, undoubtedly due to better engineering design and construction standards. Nevertheless floods can still take their toll, as shown in Figure 21 where a footbridge in Great Langdale, washed away in January 2005, has yet to be replaced.

The loss of key crossing points over major rivers could cause considerable inconvenience. In 1697 floods on the River Irthing in Cumbria led to the washing down of the bridges at Irthington and Lanercost. This was claimed to have prevented around 1,000 families from getting to market at Brampton and other towns south of river. In 1781 the river at Ravenglass in west Cumberland was so swollen with floodwater that people trying to reach Whitehaven market were delayed in getting across; they could only reach Egremont and had to sell their goods there.[28]

In the nineteenth century, before roads were surfaced with tarmac, it was not uncommon for even the main turnpikes to be gullied by floodwaters or even to be stripped to their foundations as happened to roads in the Grasmere area during the 1898 flood.[29] On the other hand floods which would have caused little difficulty to a man on horseback in the eighteenth century could easily stop a car in the twentieth (Figure 22). Even in modern times roads over upland passes like the Kirkstone between Troutbeck and Ullswater were especially liable to be damaged by flash floods.

The coming of the railways produced new sets of problems under flood conditions. Embankments could pond back floodwater and were liable to suffer from landslips and erosion. On 13 August 1891, when

22: Flooded road near Plumpton on the A6, January 2005. In the past such hazards might have been more easily negotiated on horseback. Reproduced by permission of Geoff Wilson.

serious flooding occurred throughout east Lancashire after a week of rain, an embankment between Heapey and Chorley was undercut by erosion, something which was not appreciated until some goods trucks had been derailed. Where railway lines ran through cuttings floods washing debris on to the line was a hazard.

Mills and Factories

Water-powered industrial sites, by their nature, were prime candidates for damage and destruction by floods (Figure 23). Mills normally required the construction of a dam or weir across the stream raising the water level so that it could be channeled into a lade, controlled by a sluice, which drove the waterwheel.[30] It is likely that the maintenance and periodic rebuilding or even replacement of weirs, lades and sluices was seen as an ongoing task and a regular cost for mill owners. Less frequently floods might cause damage to the waterwheels and even mill and factory buildings. In 1771 entire mills were washed away at Bolton near Appleby and Botcherby outside Carlisle, while in 1809 a woollen manufactory at Carlisle had its machinery swept away. In December 1834 a flood on the Lune caused so much damage to a large factory at Halton that it had to be entirely rebuilt.[31] In 1866 in Wigan there were worries that floods in the town might burst into old coal mines and, through them, flood modern workings; miners were evacuated as a precaution.[32] The example of the worsted spinning mill which was built at Dolphinholme in 1784 is instructive. Timothy Cragg of Ortner's diary

Table 2. Damage to Dolphinholme Mill recorded in Cragg of Ortner's diary.[33]

25/7/1787	The weir was burst
10/8/1787	The weir was washed out
9/12/1787	There was further damage to the weir
22/8/1793	The weir was breached and the water wheel broken
8/1/1796	The weir was not yet ready – a new one was being built
21/9/1796	The weir was badly damaged in a flood
29/9/1796	Over 20 men spent 2–3 days repairing the weir
14/8/1797	The weir was washed away entirely and a great number of hands were employed making a temporary replacement.

23: Old mill in Ambleside showing how vulnerable such structures were to flood damage. Photograph Ian Whyte.

records the above flood-related events in the history of the mill. The frequent interruption of production due to flood damage cannot have helped the company.

As water power sites increased in size in the nineteenth century the consequences of flood damage to the local economy could be considerable (Figures 24 & 25). A major flood in an urban area could put hundreds, even thousands, of workers out of a job (without pay) for several days and, when mills had been badly damaged, even longer. It was a matter for approving comment that during the 1927 Fleetwood flood the railway company continued to pay its workers even though it was impossible to run trains for several days.

The events associated with floods entered into local traditions. In one sense popular memories of flooding were relatively short. So many floods are described as 'the worst in living memory', that this period in practice seems to have been only 20 years or so.[34] But good stories stuck in the memory nevertheless. Timothy Cragg describes how a flood on the Lune and River Conder near Lancaster in 1793 was remembered as 'Simpson's Flood' from a man on horseback who was crossing the bridge over the Conder at Galgate when the structure collapsed. Fortunately man and horse were rescued without harm. Tantalisingly, Cragg also mentioned that other stories were told about this flood

which he could not 'well credit' and so did not record.

Newspaper reports sometimes indicate popular beliefs about floods which had no basis in reality and had something of the character of modern 'urban myths'. In Kendal there were periodic fears about the dam of the Kentmere reservoir being breached in a flood even though it collected water from only a tiny part of the catchment. It was also thought that high tides in Morecambe Bay aggravated flooding in Kendal by backing up the Kent, even though the river at Kendal was over 100 feet above sea level. Classic yardsticks of flood severity included the stopping of the Windermere ferry because the lake had risen and submerged the landing piers, and the joining together of separate lakes like Rydal Water and Grasmere, Brothers Water and Ullswater and Derwentwater and Bassenthwaite.

24: Brewery located right beside the River Caldew, Carlisle. Photograph Ian Whyte.

Costs of flooding

Flooding in Britain today is expensive, not only in terms of the damage it causes but also with the high costs of flood protection. Five years ago only £332m a year was being invested in flood defences in England and Wales; in 2005 it rose to £564m, mostly spent via the Environment Agency. Now the Agency is asking for £1 billion. The total bill for damage in the floods of autumn 2000 was £1.3 billion. £860m for domestic and £440m for commercial property. The damage in Carlisle after January 2005 ran to at least £500m, and the costs of the damage caused by floods in summer 2007 around £6 billion. The direct, tangible damage caused by flooding is comparatively easy to measure. More difficult to assess are indirect losses such as impacts on insurance premiums, house prices or tourist income in affected areas or the effects on the physical and mental health of flood victims.

Floods often affect most severely those who are least able to cope with them. In the countryside houses on floodplains are often more expensive than average because, until recently at any rate, an attractive view of the

25: Part of the former gunpowder works at Elterwater which was badly damaged by a flood in 1898. Photograph Ian Whyte.

river was a selling point rather than a drawback. In the north west, however, much of the housing which is most likely to be flooded is terraced property close to former mills and factories and the rivers which powered them. People living in these areas are less likely to have their house contents insured than elsewhere. There is also a marked tendency for areas liable to flooding from the sea to have lower house prices with concentrations of modest terraced housing in places like Blackpool, Cleveleys, Fleetwood and Morecambe, as well perhaps as a negative effect on house prices caused by greater realisation of the scale of danger.[35]

Table 3 shows that the costs of individual floods in the past are hard to measure and that comparisons over time are even more difficult because of problems of comparability. In recent years the figures used tend to be for total insurance claims. These do not include the costs to government agencies and local authorities of repairs to flood defences or bridges or the expense of upgrading flood protection in the future. Estimates of insurance claims, produced immediately after a disaster may well be underestimates and not every flood victim is insured. In the past the figures that were recorded tended to be for more specific items such as the replacement of a bridge. Another way of looking at flood impact is the number of properties flooded. Table 3 combines these measures with conversions to modern prices.

Floods are likely to have been less damaging in the past because people's possessions were fewer, more robust and more repairable than today. Flooded houses would have been easier to clean up than today. Floodwater in the past may also have been less contaminated with sewage.

In 1927 when Glenridding was flooded (Chapter 7) nobody in the village was insured. Even today, while building insurance may be a

Table 3. The Estimated Costs of Some Past Floods.

Date	Location	Cost at contemporary prices	Real cost in 2005 prices	No. of properties affected
1687	Arthuret	£4,000 – houses, animals, crops	£550.000	
1698	Armathwaite	£400 – bridge	£40,700	157 houses washed away, hundreds damaged
1720	Lancashire coast	£10,000 plus	£1.2 million	
1748	Temple Sowerby	£550 – bridge	£67,500	
1792	Brough	*c.* £1,000	£90,000	
1822	Appleby area	£7,000 bridges	£500,000	
1898	Kendal	£10,000	£750,000	
1927	Glenridding	£50,000	£2 million	
1953	Wansfell, Ambleside	£50,000	£933,000	
1954	Kendal			300
1964	Kendal			100+
1966	Borrowdale	£200,000	£2.5 million	
1966	Langdale	£200,000	£2.5 million	
1968	Appleby	£250,000	£6 million	
1968	Langwathby	£100,000 to replace bridge	£1.1 million	
1968	Carlisle	£500,000 not including uninsured property	£5.8 million	Nearly 600
1982	Carlisle			*c.* 400 homes and 50 industrial properties
1995	Appleby area			*c.* 200 properties
1999	Kendal	£200,000	£232,000	
2005	Carlisle	£450 million+ £30 million for new flood defences.		3,000+
	Keswick			165
	Cockermouth			148
	Appleby			53
	Penrith			35
	Eamont Bridge			35

Price comparison data from: http://www.measuringworth.com/ppoweruk/
(Accessed 14/10/08)

compulsory part of mortgage packages today, many people do not have contents insurance or, if they do, are inadequately covered against the scale of damage that can result from a major flood. Often it is the least wealthy people, without cover, who are the most vulnerable to flooding. In the recent floods in the Midlands and Yorkshire around a quarter of householders – half in some areas – were not insured.

In modern times damage has increased with growing spending on floor coverings such as wood block and fitted carpets and the growing complexity of electrical equipment whose salvage value is nil. We live in a 'throw away' society where even slightly damaged household goods are written off by insurers while in the past there was a more repair-oriented culture. As many insurance companies guarantee to replace new for old there is no incentive to salvage or repair slightly damaged items.

It is relatively straightforward to chronicle the effects of floods on human society in earlier centuries as well as in modern times. In this chapter we have seen that much of the damage caused by floods relates to the nature of land use and human activities on floodplains. In the next chapter we look more specifically at how floods can be caused, or worsened, by human agency.

Notes

[1]   F. Furedi, 'The changing meaning of disaster', *Area*, 39 (4) (2007), pp. 428–89.

[2]   CRO (K), WDX 393.

[3]   CRO (K), WDX 393; A. D. M. Harvey, 'Geomorphic effects of a 100–year storm in the Howgill Fells, Northwest England', *Zeitschrift fur Geomorphologie* 30 (1986), pp. 71–91; A. D. M. Harvey & R. C. Chiverrell, 'Carlingill, Howgill Fells', in R. C. Chiverrell, A. J. Plater & S. P. Thomas (eds.), *Quaternary of the Isle of Man and North West of England: Field Guide* (Quaternary Research Association, 2004), pp. 177–93.

[4]   CRO (K), WDX 393.

[5]   R. M. Johnson & J. Warburton, 'Flooding and geomorphic impacts in a mountain torrent. Raise Beck, Central Lake District, England', *Earth Surface Processes and Landforms* 27 (2002), pp. 945–69.

[6]   Cumbria Record Office, Carlisle (henceforth CRO (C)), Q/11/1/6/3.

[7]   CRO (K) WDY 193.

[8]   J. Beck, 'The church brief for the inundation of the Lancashire coast in 1720', *Transactions Historic Society of Lancashire and Cheshire* 105 (1953), pp. 91–105.

[9]   T. J. Marsh, Sand pilots: a study of the history and chronology of the guides to Morecambe Bay Sands 1501–2006, Lancaster University MA Dissertation D/5912, 2006.

[10] *Westmorland Gazette* 26 Aug. 1966.

[11] *Westmorland Gazette* 25 Nov. 1893.

[12] *The Times* 26 Dec. 1794.

[13] CRO (C), Q/11/1/100/13.

[14] CRO (C), Q/11/1/263.

[15] H. S. Cowper, *The Oldest Register of the Parish of Hawkshead in Lancashire 1568–1704* (Bemrose, 1897), pp. lx–lxi.

[16] Wyre Borough Council, Policy *Statement on Flood and Coastal Defence* (2000), p. 4.

[17] Environment Agency, *Eden Catchment Flood Management Plan* (2005), p. 36.

[18] B. P. Hindle, *Roads and Trackways of the Lake District* (Moorland, 1984), p. 154.

[19] E. Garnett, *The Wray Flood of 1967* (Centre for North-West Regional Studies, 2002).

20 P. Bicknell (ed.), *The Illustrated Wordsworth's Guide to the Lakes* (Select Editions, 1984), pp. 119–20.

21 A. B. Appleby, *Famine in Tudor and Stuart England* (University of Liverpool Press, 1978).

22 LRO, DP376/2.

23 LRO, UDWd 69.

24 S. Denyer, *Traditional Buildings and Life in the Lake District* (Gollancz, 1991), pp. 95–102.

25 R. Watson & M. M. McClintock, *Traditional Houses of the Fylde* (Centre for North-West Regional Studies, 1979).

26 LRO, QSP/1544/7.

27 CRO (K), WQ/SR/213/8, 215/23, 220/11, 213/19.

28 *Cumberland Pacquet* 1 Jan. 1782.

29 CRO (K), WDX 393.

30 S. Downward, & J. Skinner, 'Working rivers; the geomorphological legacy of English freshwater mills', *Area* 37 (2) (2005), pp. 138–47.

31 Anon., *Lancaster Records or Leaves from Local History 1801–50* (Lancaster, 1868).

32 *The Times* 19 Nov. 1866.

33 LRO, DDX 760/1.

34 Bayliss & Reed, Use of Historical Data.

35 G. Walker, J. Fairbairn, G. Smith & G. Mitchell, *Environmental Quality and Social Deprivation, Phase II. National Analysis of IPCC, Flood Hazard and Air Quality* (Environment Agency, 2003).

CHAPTER 5

Human Impacts on flooding

While floods might seem superficially to be purely natural disasters their effects can be greatly enhanced (and also sometimes reduced) by a wide range of human activities. Hydrologists do not always agree on how important these have been, but there is a widespread belief that the impact of modern floods has been worsened by human interference. Land use changes, especially agricultural improvement and drainage, have frequently been blamed for accelerating runoff into rivers and streams, not just in recent times but as far back as the seventeenth century. In the late eighteenth and nineteenth century extensive areas of upland commons in the Pennines and Lake District were enclosed under acts of parliament. In physical terms this involved walling the allotments which were granted to former commoners, building access roads with culverts and drains, and improving the soil by better drainage. During the French wars between 1793 and 1815 a good deal of this land was ploughed up, albeit temporarily.[1] It is likely, however, that any impacts on flooding were purely local though in 1668 flood damage at Tottington near Bolton was blamed on new drainage ditches which had been dug when the common was enclosed.[2] In the 1840s the development of machinery which could produce cheap tile drainage pipes to a standard bore led to a boom in the undersoil drainage of arable land and pasture. This was widely suspected by contemporaries to be a cause of more frequent flooding. Since the Second World War many moorlands have been dried out by the cutting of drainage 'grips' or surface channels in the peat. These too have been thought to have caused quicker floods and possibly higher peaks. More generally in the uplands the steady increase in sheep numbers since the Second World War, at least until the foot and mouth outbreak in 2001, is thought to have increased the flood risk by reducing the vegetation cover. Heather moorland has been replaced by acid grassland and upland soils have been compacted so that they absorb less water.[3]

Afforestation, particularly the planting of large blocks of conifers by the Forestry Commission from 1919 onwards, is a more ambivalent influence on flooding (Figure 26). It has been shown to increase the flood risk when the trees are young because of the deep drainage ditches between them can carry off rainwater rapidly. On the other hand the trees reduce the likelihood of flooding when they are older because their roots absorb more water.[4]

Mining and quarrying have sometimes been blamed for causing floods. Material containing heavy metals derived from upland mining waste can be detected in river sediments in lowland areas from medieval times onwards and may have affected livestock. There have been suggestions that the extinction of char in Ullswater was due to the inwash of heavy metals from the Greenside mines by floods. However, while mining waste has choked some specific valleys it is unlikely to have been a major cause of flooding.

As we have seen, bridges have been important in improving communications within the north west since medieval times. Bridge piers and arches constrict river channels, encouraging water to build up above them in times of flood, a tendency which is worsened by debris being piled up against the bridge. In the nineteenth century the construction of railway embankments across floodplains could also have the effect of ponding back floodwater. The modern equivalent is motorways with wide embankments whose hard surfaces and efficient drainage systems can shed water into streams much faster than ordinary soil.

26: Plantations of conifers around Thirlmere were originally intended to counteract the effects of floods by binding the soil and reducing erosion. Photograph Ian Whyte.

An even greater cause of flooding was posed by the existence of water power sites where streams were dammed with weirs and the water diverted to mill races driving water wheels (Figure 27). Manorial grain and fulling mills are recorded in increasing numbers from the twelfth and thirteenth centuries. Until the eighteenth century most of these sites were small. In the southern Lake District and the valleys of the Pennines in particular there were often strings of mills for a variety of purposes, all harnessing the same water supply. The upper Kentmere valley near Kendal had 24 water-powered sites in use at different times between the thirteenth century and 1970.[5] The holding back of water by weirs was the cause of a great deal of localised flooding in the past and led to frequent

27: A weir on the
River Kent at
Staveley. Photograph
Ian Whyte.

damage not just to weirs but sometimes to the mills with which they
were associated. The number and size of water-powered sites increased
from the late eighteenth century with woollen, flax, cotton, silk,
gunpowder and bobbin mills adding to the range of functions.

The drainage and reclamation of land from valley meadows, peat
mosses or coastal salt marshes has had a double effect in increasing the
flood hazard in the north west. Draining areas of land which formerly
acted as washlands by absorbing flood water and reducing the severity of
flooding elsewhere increased the flood risk downstream.[6] At the same
time the protection of reclaimed land by embankments was a hostage to
fortune. Such defences in coastal areas were vulnerable to overtopping by
both rivers and the sea. Once a bank was breached the defences then
reduced the speed at which floodwater could be drained back off the
land.

The hard, impermeable surfaces of urban areas reduce the amount of
rainfall absorbed by the soil and increase the speed at which it gets into
streams and rivers, making floods build up faster and raising their peak
discharges. In the nineteenth century the growth of streets of housing
often led to new sets of drains being connected to existing main drains
which were not enlarged accordingly, leading to a tendency for them to
become full after sudden rainfall. Even in rural areas motorways can have
the same effect. A severe flood at Tebay beside the M6 in 1976 was
blamed on insufficient culverts put in to drain water through the
motorway embankment. The construction of culverts, whether small
ones to carry field drains under a road, or large ones carrying a stream
through a town centre, can cause problems when the volume of flow of
the stream exceeds the capacity of the pipe or blocks the entrance.

Penrith has a long history of culverted streams such as the Dog Beck and Thacka Beck backing up during floods.

But building on river floodplains is the most widespread and common way in which people have increased the flood hazard, from the eighteenth century onwards, especially since 1945. In settlements like Burnley, Nelson and Colne there is almost continuous linear development along the main river valleys, reflecting the former importance of the stream as sources of water power, with lots of terraced housing close to the mills and factories. In such areas many of the rivers have had their channels modified, culverted or contained by retaining walls. Floodplain development may be done on the basis of underestimating the flood risk, overestimating the effectiveness of flood protection measures, not considering the risk of flooding at all, or knowing about it and just ignoring it. In the seventeenth century in a town like Kendal there were very few structures on the floodplain at all; the parish church, a bridge or two and the town's mill but no other industries or housing. So when floods did occur, as in 1635 and 1671, only the parish church and rectory was affected and a bonus was left by the retreating waters in the form of fish which were trapped inside the churchyard wall.[7]

Set against all these negative influences the building of reservoirs for domestic water supply or industrial use has undoubtedly reduced the flood risk in many parts of the north west by regulating streamflow. The mid and later nineteenth century saw a wave of reservoir building in the Pennines and the Lake District. The Thirlmere reservoir was opened in 1894 but many smaller reservoirs in the Pennines date from the 1840s and 1850s. The ones at Rivington, on the edge of the west Lancashire moors, for example, were built from 1857 onwards to supply water to Liverpool. By the end of the century many upland streams had their flows controlled in this way. Even when floods fill a reservoir to full capacity the flood peak downstream can be reduced by releasing the water more gradually.

Given the far greater amount of information on flood risk which is available compared with earlier times, it might be reasonable to suppose that new developments would either avoid flood-prone areas altogether or at least put adequate protection in place. However, the planning system often fails in both respects and new housing estates and other types of developments can be approved in manifestly unsuitable areas.

This can be demonstrated by the example of Lowther Park and Silverdale Drive in Kendal (Figure 28). This area, a kilometre east of the River Kent but crossed by the Stock Beck, was well known to local residents as being liable to flood in winter. Planning permission for the development of 49 dwellings was given by Westmorland County Council in 1972. In 1976, however, it was considered that no planning permission

28: Houses in
Lowther Park,
Kendal. Photograph
Ian Whyte.

should be given in the Stock Beck area unless developers agreed to make adequate arrangements for the disposal of surplus water. Following flooding in the early 1980s the South Lakeland District Council (SDLC) refused several requests for planning permission for these reasons. In 1986 a report on drainage in the area considered it totally inadequate to deal with a once in 50–year flood. The situation might have improved if the £3.66 million Stock Beck drainage scheme, approved in 1990, had been carried out. It incorporated plans for a flood storage reservoir and a series of new culverts. However, the scheme was deferred due to lack of money.

In 1989 plans for residential developments in this area were eventually approved, The SLDC had opposed the scheme but were told that they could not hold it up unless they could prove evidence of a flood risk. Department of the Environment inspectors would not accept local knowledge of flooding as admissable evidence and the SLDC could not afford to fight repeated appeals. So the scheme went through, with part of the development being undertaken by Two Castles Housing Association and part by Lowther and Dawson. In 1993, four years after the houses were completed, a flood caused partly by melting snow on the fells reached the doorsteps of the new houses. In January 1999, after four inches of rain in 24 hours, the houses were badly affected by flooding, necessitating the evacuation of around 100 people. Damage to the fabric and content of the houses was up to £20,000 each. The SLDC's acting chief executive called for a full investigation and the council's head of environmental protection immediately confirmed the residents' views that the drainage system in the area was inadequate. The builder pointed out that both the surface water drains and sewerage systems had been adopted by the local authority without complaint. The residents, scared that the prices of their properties would fall and their insurance premiums rise, and that the foundations of their houses might have been undermined, formed an action group. It was pointed out that because of the wetness of the site the surface water sewer ran at 50 per cent capacity even under normal conditions, leaving little leeway to cope with emergency flows.

Following the floods the council made the Stock Beck Flood Alleviation Scheme a priority and applied to DEFRA for assistance. In February 2005 the area was nearly flooded again. The £3.3 million scheme was eventually finished in 2007, protecting *c.* 170 homes by the construction of a flood storage reservoir and the replacement of an inadequate culvert (Figure 29).

We have now looked at the causes of floods in north west England and have seen how complex they can be. In the next chapter we review the flood histories of two very different catchments within the region.

29: Area of the flood storage reservoir on the Stock Beck, Kendal. Photograph Ian Whyte.

Notes

1   I. D. Whyte, *Transforming Fell and Valley. Landscape and Parliamentary Enclosure in North West England* (Centre for North-West Regional Studies, 2003).

2   LRO, QSP/316/1.

3   H. G. Orr & P. Carling, 'Hydro-climatic and land use changes in the River Lune catchment, North West England. Implications for catchment management', *River Research and Applications* 22 (2006), pp. 239–55; J. E. Roe & A. Parker, 'Techniques for validating the historic record of lake sediments. A demonstration of their use in the English Lake District', *Applied Geochemistry* 11 (1996), pp. 211–15.

4   P. Carling & K. Beven, 'The hydrology and geomorphological implications of floods: an overview', in P. Carling & K. Beven (eds.), *Floods. Hydrological, Sedimentological and Geomorphological Implication* (Wiley, 1980), pp. 1–9; Archer, *Land of Singing Waters*, pp. 124–5.

5   J. Somervell, *Water-powered mills of South Westmorland* (Titus Wilson, 1930).

6   Orr & Carling, Land use changes.

7   J. F. Curwen, *Kirkbie Kendall* (Titus Wilson, 1900), pp. 2, 21.

CHAPTER 6

Case Studies of Flooding. The River Eden: a rural catchment. The River Irwell: an urbanised and industrial catchment

Something of the variety of the human experience of flooding within north west England can be appreciated by looking at the flood histories of two contrasting river basins, the River Eden in Cumbria and the River Irwell which separates Manchester and Salford. The former is largely rural apart from the city of Carlisle while the latter has a long history of industrial and urban development.

The River Eden

The Eden is one of the largest rivers in northern England (Figure 30). It rises at the head of the Mallerstang valley and flows for some 90 miles to the Solway, receiving water from short tributaries draining the steep Cross Fell escarpment and longer ones like the Caldew, Eamont, Lowther and Petteril from the west. The central part of the Eden valley lies in the rain shadow of the Lake District and so has moderate rainfall but the hills around it have much higher totals. The Eden falls rapidly from a height of 690 metres at its source to 160 metres at Kirkby Stephen and thereafter drops more gently. The greatest flood danger in the lower part of the valley occurs when floods on the main river peak at the same time as ones on the principal tributaries. Ullswater and the Haweswater reservoir reduce flooding from part of the catchment. In terms of sources of flooding 50 per cent comes from the main River Eden, 17 per cent from tributaries, 25 per cent from surface water drains and 8 per cent from sewers.[1] The Eden has some good-quality agricultural land developed from glacial tills overlying sandstone. Settlement in the valley is often in the form of large nucleated villages, some of which have streams flowing right through their central greens, as at Maulds Meaburn. Major floods tend to occur between January and March due to prolonged frontal rain brought by Atlantic depressions, sometimes augmented by melting snow

30: The catchment of the River Eden. Map drawn by Simon Chew, Lancaster Environment Centre.

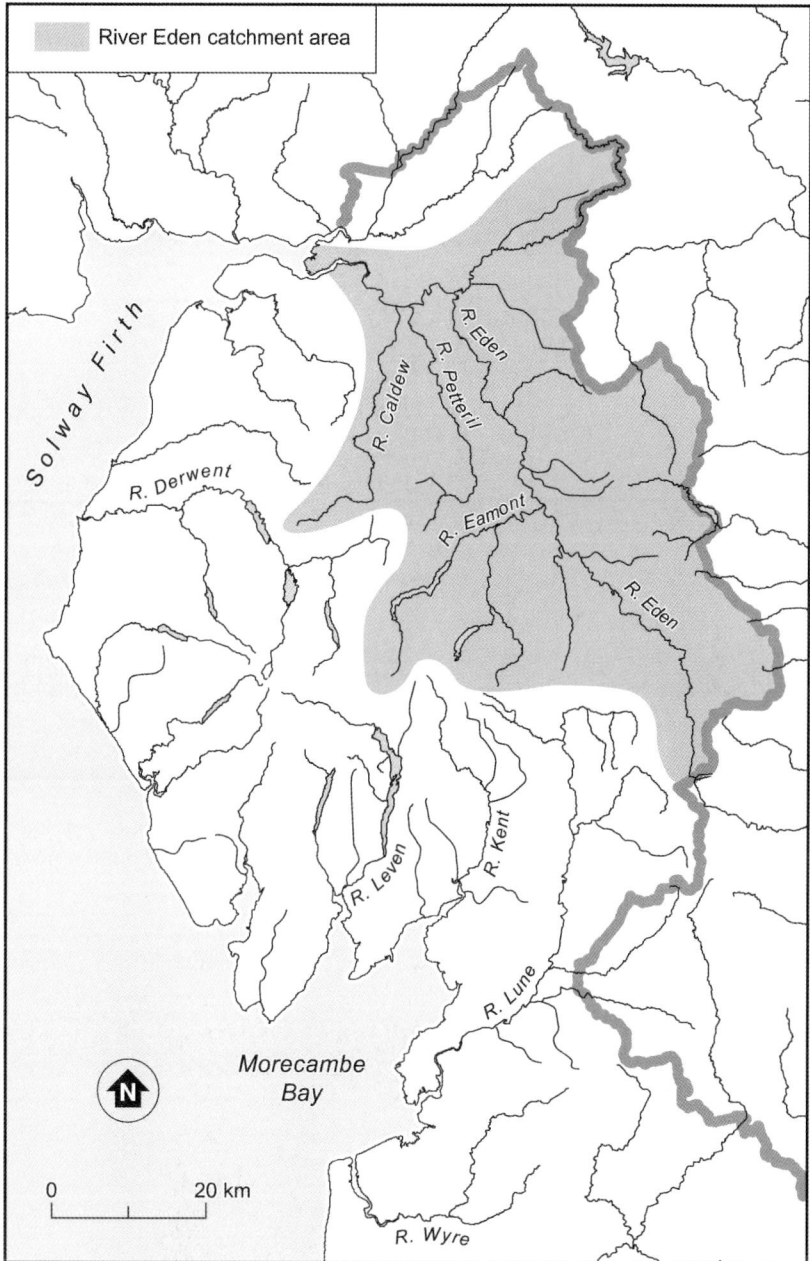

River Eden catchment area

Solway Firth

R. Eden

R. Petteril

R. Caldew

R. Derwent

R. Eamont

R. Eden

R. Leven

R. Kent

R. Lune

Morecambe Bay

N

0   20 km

R. Wyre

on the surrounding fells. Because of the length of the catchment there is a distinct time lag between flooding in the upper reaches and floodwater reaching Carlisle. The valley has long been a major through route; to the south via Shap and the Lune Gorge, over Stainmore into Durham and Yorkshire, east through the Tyne gap into Northumberland and north to Scotland. Bridges on these main routes have been important for centuries.

The towns in the Eden basin, like Kirkby Stephen and Penrith, are mostly small rural market centres. Even Carlisle had a population of only 71,773 in 2001. Carlisle is situated on three rivers – the Eden, Caldew and Petteril. The medieval core of Carlisle ran from the castle south eastwards past the cathedral to the English Gate, between, but well above, the rivers Caldew and Petteril. There were only small suburbs on the west side of the Caldew at Caldewgate, and at Rickersgate, on the floodplain between the city walls and the bridge leading north to Scotland. The growth of the city from the mid eighteenth century was predominantly on the areas of land known as 'holmes', the low-lying ground within the meanders of the River Eden, including the filled-in south channel of the river, supposedly abandoned due to a change in course resulting from an especially high flood in 1571. Expansion of both residential and industrial areas on to the floodplains led to a growing number of properties at risk. In 1801 Caldewgate had a population of 1,990 but by 1821 this had grown to 3,915. The low-lying suburbs of Botchergate and Rickersgate grew at comparable rates. The removal of the city walls in 1821 and the filling in of the southern channel of the Eden encouraged growth along the riverbanks. By 1901 Caldewgate had become a mixed area of factories and tightly-packed housing. Shaddongate had spread south to Denton Holme and there had also been growth to the east along Warwick Road and London Road, and around Brunton Park and Botcherby. The Willow Holme Industrial Estate downstream from the Castle was planned in the late 1950s. Construction, in this area, known to have flooded several times in the past, started in 1960, and most sites were occupied by 1965, just in time for the major flood of 1968.[2]

Carlisle has suffered in the past because of its location. It is vulnerable not only to floods from the Eden but also from the Petteril and Caldew. Both of these drain different areas so they do not necessarily all flood at the same time, or at the same level of intensity, but when they do then the city has problems.

The medieval borough of Appleby is located in a bend of the Eden rising steeply from the parish church of St Lawrence at the bottom of the main street to the Castle at the top, from which the ground falls in steep bluffs to the river. The most vulnerable areas for flooding lie around the church and in the small suburb known as The Sands, to the west of the river (Figure 31). Penrith, located well away from the Eden or even major tributaries like the Eamont, might seem to be in a less flood-prone location than other Eden valley towns. However, it has a long history of localised flooding from smaller streams which run in culverts under parts of the town.

In the flood of November 1771 which caused the famous bog burst on Solway Moss, the Eden supposedly reached a higher level than ever previously recorded, high enough in Rickersgate (Carlisle) to drown a

31: The Sands at Appleby, the part of the town most vulnerable to flooding. Photograph Ian Whyte.

horse in its stable. In Appleby the river damaged the new jail or 'house of correction' and flooded the churchyard 18 inches deep. Furniture was afloat in houses in The Sands and two of the arches in St Lawrence church subsided.[3] On 25 August 1792 another flood occurred on Carlisle's fair day. The fairground was under water and livestock had to be ferried across the swollen river from Stanwix to a temporary site in the city.[4]

The worst flood in historic times in the Eden basin, before 2005, occurred between 1 and 3 February 1822. In Carlisle the embankment between the racecourse and the cattle market gave way suddenly flooding Rickersgate early on a Sunday morning when many people were still in bed. Several dragoon horses were nearly drowned in their stables and the water reached the ceilings of many houses in the lowest lying area of the suburb. Many Rickersgate shopkeepers were ruined and some ladies in the town started their own relief fund to provide food and coal to the victims. The bridge at Botcherby (only built in 1817) and one at Harraby, were badly damaged. There was extensive flooding in both Penrith and Appleby but the disaster of 1822 was notable for the number of bridges which were completely washed away in the upper part of the catchment. Bridges at Appleby, Asby, Blandswath, Bolton, Brough, Brougham, Cliburn, Eastwath, Great Musgrave, Hoff, Kirkby Stephen, Kirkby Thore, Long Marton, Soulby, and Temple Sowerby were washed down or badly damaged and the county authorities had a repair bill of unprecedented size.[5]

The nineteenth century saw a number of serious floods in the Eden valley caused by rapidly melting snow including February 8–9 1831 and 8 December 1856. The latter flood was particularly bad in Kirkby Stephen where about a quarter of the houses in the town were affected and a brewery was badly damaged. February 1831 had been one of the worst floods in living memory in Penrith but the town also suffered on 10 December 1852 when a culvert near the new brewery backed up flooding several nearby streets. On 25–26 November 1861 the Dog Beck in Penrith flooded, something which the locals blamed on recent drainage on the higher land above.[6]

Most of the really bad Eden valley floods of the nineteenth century were winter ones but not all of them. On 5 July 1881 a summer flood swept away a railway embankment near Carlisle leaving the tracks hanging in mid air. On 24 July 1888 a thunderstorm caused a severe flash flood at the head of the River Raven above Kirkoswald which washed away several acres of peat. Trees alongside the river were uprooted all the way down to Kirkoswald and hayfields were buried by sand and gravel. On 9 August 1894 a similar flash flood in Geltsdale swept away the parapet of a bridge and deposited the stones seven miles downstream. After the waters had subsided trout were found in the fields 500 yards from the stream.[7] The worst flood on record in the southern Lake District occurred on 2 November 1898, but conditions were also bad in neighbouring parts of the Eden catchment. There was a great deal of damage to walls and roads in Patterdale and in nearby Boredale hundreds of cartloads of sand and gravel were washed down on to the roads.[8]

The early twentieth century saw fewer really severe floods but on 2 January 1925 melting snow caused extensive flooding in the western part of Carlisle. A relief corps was organised to bring soup and other food to people trapped in the upper rooms of their houses; the mayor went round personally to identify the worst cases of distress. This was one of the first floods in Carlisle where damage to electrical fittings is mentioned. A number of factories and a new electricity station on Willow Holme were damaged while Carrs' biscuit factory was several feet deep in floodwater. There was a temporary shortage of bread as the Carlisle Bread and Flour Company's ovens were under water. In Penrith the culverts of the Dog and Thacka Becks again proved to be too small to cope with the floodwater and overflowed.[9]

On 24 March 1968 one of the worst floods in modern times occurred with around 400 properties in Carlisle being flooded as the Caldew and Petteril overflowed as well as the Eden. The changes in people's lifestyles and material possessions since the end of the Second World War were highlighted by descriptions of damage to washing machines, fridges and telephones, and by the £30,000 worth of damage at Mackenzie's Motors where many new cars were submerged. Further up the Eden the most notable incident was the washing away of the bridge at Langwathby (Figure 32).[10]

The floods of January 2005 were exceptionally severe with an estimated 200–year return period. At the time of the disaster proposals for new flood defences were being displayed for public consultation in

32: The bridge over the River Eden at Langwathby built to replace the one destroyed by floods in 1968. Reproduced by permission of Geoff Wilson.

the Environment Agency's offices in Penrith. According to normal procedure they were designed to protect against 100-year events so that even if they had been completed, they would have been overtopped. The scheme has been designed to defend against floods with a return period of one in 200 years i.e. a 0.5 per cent chance of flooding in any one year. However, there is always a possibility that even these defences could be overtopped: no flood defences can be considered to be completely safe.

The River Mersey and River Irwell

The River Mersey is formed by the junction of the Rivers Tame and Goyt at Stockport. Its floodplain cuts a wide swathe of green country from east to west between the suburbs of Manchester from Stockport to Urmston before joining the Manchester Ship Canal. The River Irwell rises north of Manchester, flowing through Ramsbottom, Bury and the outskirts of Bolton into Manchester itself, dividing the city proper from Salford before becoming, in its lower reaches, the Manchester Ship Canal. These rivers, and tributaries like the Rivers Irk and Medlock which flow through the heart of Manchester, provide a major flood risk for a substantial population. As early as 1750 it was noted that in winter and often too in summer floods made the Mersey impassable leading to loss of life and delays in the postal service. The risk of flooding was greatly increased in the later eighteenth and nineteenth centuries by the spread of industry and the building of working-class housing close to riverside factories.

The earliest flood we have identified for the Irwell was in 1616 and a number of other seventeenth-century inundations are recorded. A flood on 8 October 1767 was the worst anyone remembered with the Irwell and Mersey both bursting their banks, carrying away large quantities of hay and corn and causing several hundred pounds worth of damage to properties in Salford though it is notable that even at this date the damage described was still largely rural.[11]

A flood on 17 August 1799 was graphically described by Mrs. G. L. Banks in her novel *The Manchester Man* (1876):

> The tannery yard, high as it was above the bed of the Irk, and solid as was its embankment, was threatened with invasion … the surging waters roared and beat against its masonry … Men with thick clogs and hide-bound legs, leathern gloves and aprons were hurrying to and fro with barrows and bark-boxes, for the reception of the valuable hides which their mates, armed with long-shafted hooks and tongs, were dragging from the pits pell mell ere the advancing waters should encroach upon their territory and empty the tan pits for them…

> Already the insatiate flood bore testimony of its ruthless greed. Hanks of yarn, pieces of calico, hay, uptorn bushes, planks, chairs, boxes, dog kennels and hen coops, a shattered chest of drawers, pots and pans had swept past, swirling and eddying in the flood, which by this time spread like a vast lake.

> Too busy were the tanners, under the eye of their master, to stretch out hand or hook to arrest the progress of either furniture or live-stock though bee hives and hen coops and more than one squealing pig went racing with the current.[12]

One of the first floods for which we have a good deal of detail occurred on 30 December 1815 due to rapid melting of snow on the Pennines. The Irwell rose three feet higher than the previous record level. Haystacks, timber, pigs, poultry and even a large iron boiler from a steam engine were seen being carried downstream through Manchester. Damage to farms and factories was estimated at many thousands of pounds. A barge broke loose from its moorings and sank after being driven into Regent Bridge near Hulme. Another barge which was swept away was left high and dry when the floodwaters dropped. Part of the Black Boy public house in the old Church Yard collapsed into the Irwell, which had undermined its foundations; fortunately no one was hurt.[13]

On December 21 1837 there was another serious flood in which, as well as livestock and furniture, a baby in a cradle was seen being swept down the Irwell.[14] Another significant flood on the Irwell on 1 November 1843 washed away some staging at Hunts Bank which had been erected for the construction of a railway bridge. It floated down the river as a raft and carried away one of the uprights of a bridge causing the structure to collapse.[15] A flood in May 1847 caused a great deal of damage to property along the River Medlock, a tributary of the Irwell. The problem seems to have been caused in part by a failure to open sluice gates which provided water for local factories when river levels were low. People living in nearby courts and alleys had six or seven feet of water in their houses and had to be evacuated in boats. Police, trying to co-ordinate relief efforts, were hampered by hundreds of spectators.[16]

The Bridgewater Canal from Manchester to Runcorn crosses the Mersey on an aqueduct built in 1760 which backed up the water in the river, threatening to undermine the canal embankment so in 1840 a diversion weir was built with an overflow channel to take the floodwater under the canal at another location.

The worst flood of the nineteenth century occurred between the 15 and 17 of November 1866 when the Irwell reached 25 feet above its normal level. Trees, torn up by the roots, floated past the cathedral. The water reached above the highest arch of the new bridge between Manchester and Salford. In the Broughton and Strangeways areas many warehouses and homes were flooded. Bolton and Bury were affected as well. Hundreds of people were taken to Salford workhouse, the town hall and the Catholic Institute Dispensary for emergency accommodation. A gentleman who had rescued his wife was then swept away and drowned. The driver of a horse and cart trying to cross a bridge near the Assize Courts suffered the same fate while a man in Lower Broughton, wading through the water along a submerged footpath, fell into a hole and was drowned.[17]

A flood on 13 July 1872, due to a summer depression, affected a wider area than the inundation of 1866. Peel Park and the racecourse were flooded but otherwise there was little damage along the Irwell. It was a different story on the Medlock around Clayton Bridge where weirs and embankments were destroyed and there was much damage to local textile factories. The flood reached Manchester City Cemetery near Philips Park. In a macabre scene coffins were washed out of the soil and broken against the weir of an adjoining printworks so that corpses were swept downstream; more than 50 were later recovered. The floodwater was 15 feet deep in places and damage to houses and industrial premises across the city was extensive.[18]

The growth of the built up area of Manchester and its satellite towns has modified runoff to the Rivers Mersey and Irwell producing changes in flood patterns which have required expensive flood protection measures including embankments along the Mersey from Stockport to near Urmston. There is temporary storage for floodwater in the Didsbury flood basin, an area comprising a golf course, allotments and playing fields. When the Mersey is in flood, gates on the weir at Didsbury can be opened to let water into the basin and into the Sale and Chorlton water parks.[19]

In central Manchester the area known as Little Ireland in the streets behind Oxford Road station, close to the River Medlock and the Rochdale Canal, experienced rapid industrial growth in the early nineteenth century as the district benefited from canal transport and water for industrial purposes and waste disposal. By the 1820s this area had textile mills, iron foundries, food processing factories, slaughterhouses, coal and timber yards as well as unplanned residential housing put up by speculative builders. The result was damp housing which was highly vulnerabile to flooding. In May 1847 boats had to be used to rescue people from the upper storeys of their homes. The area is described by Frederich Engels in his *Condition of the English Working Classes*. The book contains a graphic description of an Irish family living in a cellar dwelling who, when Engels peered inside, were sitting on their door, propped up on piles of bricks, while 18 inches of water thick with sewage swirled around below them.[20]

In Lower Kelsall near Salford there were severe floods in 1866, 1946, 1954 and 1980. A large council estate was built there in the 1930s and many of its houses were badly flooded in 1946 and 1980, some with three feet of water in them. A one in 100–year flood there would cause much damage. Two flood storage basins designed to protect *c.* 3,000 homes against a one in 75–year flood have been built at a cost of £11m.[21]

In modern times the flow of the Mersey has been affected by the construction of the M60 which takes advantage of the open corridor created by the Mersey floodplain and which crosses the river several times. The blocking effect of the motorway embankments has necessitated the building of a system of storm drains to avoid the ponding back of floodwater. On the other hand gravel pits at Chorlton and Sale which were created by the removal of aggregate for building the motorway have been turned into recreational lakes which also act as flood storage areas.

The River Irwell provided a major problem due to its proximity to some of the most densely-packed residential districts in the heart of nineteenth century Manchester (Figure 33). In particular its bed was raised so much by the dumping of rubbish that sewers and mill drains became blocked. In the 1860s it was estimated that 75,000 tons of factory cinders a year were being dropped into the river. It is not surprising then to find that 29 major floods are recorded on the Irwell between 1816 and 1970 – an average of one every 15 years.[22]

Following the flood of 1866 ambitious plans were proposed to alleviate the flood danger, including by the construction of a major flood relief

33: The River Irwell
in central
Manchester.
Photograph Ian
Whyte.

tunnel from near Peel Park to Regent Bridge which, at an estimated cost of £40,000, would have accommodated half the volume of a flood on the scale of that of 1866. However, very little was done.[23]

The worst modern flood on the Irwell was in 1946 when 243 hectares were flooded in Lower Kelsall, Charlestown and Lower Broughton affecting around 5,000 homes and 300 industrial properties. Today about 10,000 properties in Salford have been classified as having a high risk of flooding. Lower Kelsall, in a meander loop of the Irwell, and Charlestown on the opposite side of the river, have some flood protection from embankments. Salford's draft flood-risk plan categorises areas as being of low risk if they are beyond the likely limit of a one in 1,000–year flood, high risk if they are within the once in a century limit and medium risk if they lie in between. New development should, as far as possible, be in the low-risk areas. Offices and other developments in the high risk area should have their lowest floors above the predicted flood levels. Basement levels should be used for car parking and new developments should in no way increase the risk of surface water drainage flooding.

This chapter has presented some flood stories from two very different river basins. However, many of the most dramatic and damaging floods in the north west occurred in small upland catchments, sometimes with devastating effects for the inhabitants, as we will see in the next chapter.

Notes

[1]   Environment Agency, *Eden Catchment Flood Management Plan Consultation Report* (2005), pp. 3–4.

[2]   K. Smith & G. Tobin, *Human Adjustment to the Flood Hazard* (Longman, 1979), pp. 74–76, 84–90.

[3]   CRO (K), WG/SD/369/14; Anon. *The Westmorland book,* vol 1 (Titus Wilson, 1888–9), p. 310.

[4]   *Cumberland Pacquet* 4 Sept. 1792.

[5]   *Carlisle Patriot* 9 Feb. 1822; Smith & Tobin, *Flood Hazard;* J. F. Curwen, *The Later Records of North Westmorland or the Barony of Appleby* (Titus Wilson, 1932).

[6]   *Cumberland Pacquet* 4 Sept. 1792.

[7]   *British Rainfall,* 1894.

[8]   *Westmorland Gazette* 5 Nov. 1898, 12 Nov. 1898; *Penrith Observer* 8 Nov. 1898.

[9]   *Carlisle Journal* 2 Jan. 1925; *The Times* 6 Jan. 1925; Smith & Tobin, *Flood Hazard.*

[10]   *Westmorland Gazette* 29 Mar. 1968; *British Rainfall,* 1968; Smith & Tobin, *Flood Hazard.*

[11]   D. W. Proctor, *Memorials of Manchester Streets* (Sutcliffe, 1874).

[12]   G. L. Banks, *The Manchester Man* (Gollancz, 1970).

[13]   *The Times* 5 Jan. 1816.

[14]   Proctor, *Manchester Streets.*

[15]   *The Times* 2 Nov. 1843.

[16]   *The Times* 11 May 1847.

[17]   *The Times* 17 & 19 Nov. 1866.

[18]   Proctor, *Manchester Streets.*

[19]   I. Douglas, 'Urban flood plains and slopes; the human impact on the environment in the built up area', in A. Gardiner, P. Hindle, J. McKendrick & C. Perkins (eds.), *Exploring Greater Manchester. A Fieldwork Guide* (Manchester Geographical Society, 1999).

[20]   F. Engels, *The Condition of the Working Class in England* (Penguin, 1987 ed.), p. 147.

[21]   Douglas, Urban flood plains.

[22]   I. Douglas, 'Geomorphology and urban development in the Manchester area', in R. H. Johnson (ed.), *The Geomorphology of North West England* (Manchester University Press, 1985), pp. 337–52.

[23]   J. Corbett, *The River Irwell* (Heywood, 1907).

CHAPTER 7

Flooding in Smaller Catchments

Floods, such as the one that hit Carlisle in January 2005, caused by long periods of cyclonic rain spread across entire river basins and even the whole region, may cause the greatest total amount of damage and distress. But it is often sudden flash floods in small upland catchments that are more dangerous and unpredictable. When a major river floods after prolonged rain there is more time to issue flood warnings, save livestock and evacuate people. A flash flood, however, can be totally unexpected and more damaging as a result. In terms of their physical impacts on the landscape they may cause more changes within an hour or two than ordinary winter river levels over several decades, leaving behind scars which may persist for many years. This is one reason why they are of such interest to hydrologists; they demonstrate the nature of the processes of erosion and deposition at a greatly accelerated rate. Floods of this kind characteristically tend to result from summer thunderstorms over fells and valley heads. Under the right conditions with rapid uplift of moist air intense falls of rain can occur.

One of the earliest floods of this kind is noted in the parish register of Hawkshead, then in north Lancashire. On 10 June 1686 a fearful thunderstorm produced a flood the like of which had not been experienced in the memory of the oldest inhabitants. Several bridges in the Hawkshead area were washed away together with a number of houses and mills. The water uprooted trees and deposited large boulders and great beds of sand in places.[1]

An even more severe episode which is documented in greater detail occurred on the fells north of Helvellyn between the Vale of St. John and Mosedale on 22 August 1749. The storm was very localised; there was no rainfall at all on Great Mell Fell, a short way to the east. The downpour, which lasted for less than two hours, was so sudden and the rainfall so intense that observers described it as a 'waterspout'. On the Mosedale side of the watershed population was thinly scattered, although several houses were washed away entirely and others were buried in sediment to the level of the first floor. The main reminder of the flood in the modern landscape of Mosedale is the spread of large stones forming a boulder bar which chokes the floor of the Mosedale Beck just above the farm of Lobbs marking where the flow of floodwater began to slacken, dropping huge rocks, many over 20 tons in weight, which would never have been shifted by ordinary levels of streamflow (Figure 34).[2]

On the other side of the ridge, in the Vale of St. John, there were more farms and cottages and the devastation was greater (Figure 35). An article published in the *Gentleman's Magazine* in 1754 included a diagram of this valley showing the steep slopes of its eastern side and the huge fans of debris brought down by the flood (Figure 36). The mill at Legburthwaite was washed away entirely as well as other houses and cottages. Several houses were flooded to first floor level. Fields on the valley floor were ruined through being buried by huge spreads of boulders and gravel. Amazingly no one was drowned although many animals were lost.[3] Interestingly the thunderstorm which burst over the fells north of Helvellyn either extended further east towards the head of Longsleddale or a separate severe storm occurred there at almost the same time. The resulting flood did not have its official chronicler but the Quarter Sessions records contain petitions for funds to repair bridges at Sadgill and Wadshaw in the valley which had been destroyed by a flood, along with hayfields and crops.[4]

34: Boulder bar, Mosedale, dropped by the 1749 flood. Photograph Ian Whyte.

The flooding of the Vale of St. John in 1749 was often linked in early Lake District guidebooks with another catastrophe a few miles to the west. On 9 September 1760, in the middle of the night, a cloudburst over the mountain of Grassmoor in the north-western Lake District led to a severe flash flood which caused considerable damage in the Vale of Lorton. The classic description is by William Gilpin, the pioneer guidebook writer, who visited the site of the flood a dozen years later, at which time the damage was still clear to see. Where the Liza Beck emerged from the mountains into the valley its normal character of 'a mere constricted rivulet' was changed into a torrent up to 18 feet deep and nearly 100 feet wide; an area of 10 acres was spread with a deep fan of debris but fortunately the farms and cottages in the area were mostly on slightly elevated sites just beyond reach of the water (Figure 37). Like the flood of 1749 there was, amazingly, no loss of life.[5]

35: The eastern side of the Vale of St. John down which the floods of 1749 swept. Photograph Ian Whyte.

36: A contemporary drawing of the 1749 flood showing the gullying of the steep valley sides and the deposition of boulders in the valley. Photograph Ian Whyte.

In Chapter 2 we mentioned the Holmfirth dam disaster of 1852. This valley also has a history of flash flooding due to summer thunderstorms. One is recorded from 1738 and a more severe one from 23 July 1777, when the stream rose 21 feet above its normal level, including a 12 foot rise in 15 minutes. This flood washed bodies out of the graveyard of Holmfirth Chapel, swept several houses away entirely, and caused major damage to roads and bridges. In addition, three men were drowned.[6]

37: Vale of Lorton. The slope of debris above the modern stream was part of the fan of debris from the 1760 flood. Photograph Ian Whyte.

On 13–14 August 1891 severe damage was caused to the mills at White Coppice, on the edge of the west Lancashire Moors near Chorley, by an unusual flood. The mill was located below a large reservoir. After heavy rain a landslide caused several hundred tons of earth to fall from the hillside into the stream above the reservoir. The debris was carried into the sluice which transferred surplus water from the Roddlesworth reservoirs to those at Rivington. The dammed-up water overflowed from the sluice and into the White Coppice Reservoir, sending a wave of water over the dam and on to the mill below. A man on duty at the mill warned the occupants of nearby houses and moved them out of danger. In the mill the water damaged much yarn and cloth while several bridges further downstream were carried away. Everyone was relieved, however, that the dam itself had not collapsed.

The steep narrow valleys around the village of Dent have a long history of severe floods (Chapter 2). One of them occurred on 9 July 1870 when a 90–minute thunderstorm at the head of the valley caused havoc at the construction sites for the Settle–Carlisle railway. A new tunnel was flooded and two workers were drowned while large quantities of building materials were washed away.[7] It is not often that the impact of pre-twentieth century flash floods are recorded in detail rather than in the sensational terms used by newspaper reporters. However, information on the flooding of the valley of Garsdale near Sedbergh on 8 August 1889 has been preserved through local relief activities under the auspices of the Garsdale Inundation Fund. A thunderstorm lasting an hour and a half burst over Grisedale, at the very head of Garsdale and sent a surge of water down into the main valley (Figure 38). Because Garsdale is narrow and steep sided many farmsteads and cottages were in vulnerable

38: This chapel in Garsdale was badly affected by the 1889 flood. Photograph Ian Whyte.

locations on the valley floor. Bridges, walls, fences and trees were washed away, hay crops ruined and many homes flooded. The school was flooded and the playground buried in debris. A committee was formed with the local vicar as secretary and an MP and a county councillor as members. They established a relief fund which raised money through private donations. The damage that each farmer and householder had suffered was inspected and valued and a report estimating the scale of everyone's losses produced. Their figures were reduced in cases where landowners had already provided help for their tenants or where the damage was irreparable, as when erosion had removed entire fields. Of the £240 raised about half was spent on repairing damage to the school and its lands. Apart from the school the greatest damage was caused to local bridges which were not maintained by the county, followed by the loss of retaining walls, drystone walling, fencing, meadows and pasture. The fund allocated its resources to each family in proportion to their needs and was then wound up. Although the subscribers included the Marquis of Ripon and the Bishop of Richmond most of the money came in small amounts from local people.[8]

The flood which caused so much damage in the village of Glenridding, near the head of Ullswater, in 1927 had one of the strangest causes and has frequently been misreported. The Greenside lead mines above the village had for a long time drawn water from a natural tarn in Keppel Cove situated below the summit of Raise, north of Helvellyn. The water in the tarn was held back by a natural barrier of glacial moraine which had only been slightly modified by the mining company which had flattened and widened its top and constructed a spillway for water to drain out over (Figure 39). The night of 28–29 October 1927 was very wet and stormy as a deep depression passed over the Lake District. Waves, built up by the wind, lapped at the outflow of the tarn and started to wash away the moraine barrier. At some point between midnight and 1 a.m. the dam began to give way. Once the water made a breach it was rapidly enlarged and within half an hour almost the entire six-acre tarn had emptied with 124,000 cubic metres of water cutting a 60-foot deep gouge in the hillside and eroding about 13,000 cubic metres of material from the dam breach.[9]

A wall of water hurtled down the valley, demolishing hundreds of yards of stone walls as it went. Stones from the breach and the levelled walls were dropped as boulder bars further down the valley. Between 1.30 and 1.40 a.m. the flood hit Glenridding three km away flooding buildings near the beck with up to two metres of water (Figure 40). The occupants of houses and cottages had to flee for their lives. Worst affected was Millcrest's Hotel (now the Glenridding Hotel) which stands right beside the stream in the centre of the village. Some of the maids who worked at the hotel were sleeping in a basement room. They only escaped with the help of Ernest Thompson, the 'boots' at the hotel, who pulled them to safety through a window. He was recommended to the Royal Humane Society for a medal. Hundreds of tons of sand and gravel as well as boulders up to six-feet high blocked roads, ruined fields and filled gardens (Figure 41). Apart from the hotel 13 houses, cottages and businesses were badly damaged and most were uninsured. The shortage of water after the disaster gave rise to worries that the lead mines would have to close through lack of power, adding to the stress of local families, many of which had members working at the mines.[10]

39: Keppel Cove (left) with breach in moraine dam in centre and later concrete dam to the right. Photograph Ian Whyte.

40: View of
Glenridding village
showing the steep
gradient down which
the flood from the
dam burst of 1927
poured. Photograph
Ian Whyte.

41: Boulder bar above
Glenridding dropped
by the 1927 flood.
Photograph Ian
Whyte.

Following the disaster the mine owners built a 40-foot high concrete dam just below the site of the natural one that had burst. Unfortunately, on the night of 19–20 August 1931 this dam also failed following two days of storms and heavy rain, causing another flood and further damage in Glenridding where furniture, stock in shops and cars were damaged and haystacks were washed into Ullswater (Figure 42). Hundreds of yards of retaining walls in the village which had been built at the expense of the mining company after the 1927 flood were washed away. Once again tons of boulders and gravel were dumped on fields, gardens and roads. The dam, with a gaping hole in it, can still be seen above the village.[11]

42: The dam above Glenridding which failed in 1931. Photograph Ian Whyte.

Flooding rarely made the headlines during the early 1940s because of more urgent war news and restrictions on reporting items which might affect morale. However, a flood caused by a thunderstorm on the borders of Lancashire and Yorkshire near Holmfirth on 29 May 1944 was exceptionally severe, including hailstones supposedly the size of mothballs. Some of the floodwater went west down Longdendale but most of it surged eastwards bursting the banks of the stream at Holmfirth and causing major damage to roads and bridges. A number of houses, shops and mill buildings were demolished or badly damaged by water up to six-feet deep. Bales of wool and cotton from flooded mills were washed through the streets. Three people were drowned and Geoffrey Riley, aged 14, was awarded a medal for trying to save a woman from drowning; his father, who had also been involved in the rescue attempt, was swept away by the current and drowned.[12]

The stream at Glenridding certainly looks as if it could mean business during a major flood but at the north end of the village of Troutbeck, near the famous Mortal Man inn, you have to look carefully for signs of the stream which brought death and destruction on June 25 1953 (Figure 43). A sudden cloudburst over Wansfell and the Kirkstone Pass washed out long stretches of road and set off landslides which blocked other sections as well as washing away bridges, trapping coaches and cars and cutting off farms and cottages. Some houses in Troutbeck were flooded to a depth of four feet. One hundred yards below the Mortal Man the Scot Beck ran under the road in a culvert. A man from Leicester who was staying at the inn tried to help save some ducks and poultry without realising that the floodwater was backing up behind a wall a short distance above him. The wall suddenly collapsed and the rush of water

43: The part of Troutbeck which was devastated by the flash flood of 1953. Photograph Ian Whyte.

swept him to his death. In Ambleside, which did not even experience the worst of the storm, hailstones the size of pennies were carried away in wheelbarrows, while scores of shops and houses were flooded. It was a week before the Kirkstone Pass road was re-opened and years before some of the scars caused by gullying began to fade.[13]

The village of Wray in the Lune valley suffered an experience similar to Lynmouth on a smaller scale on 8 August 1967 when a cloudburst over the Bowland Fells, estimated at 117mm in 90 minutes, sent a 200-foot wide wall of water down the rivers Hindburn and Roeburn. Nine houses in the village were totally destroyed and many others damaged, making 32 people homeless and prompting a national appeal for aid. The story of the Wray flood has been told in detail by Emmeline Garnett in another book in this series,[14] but it is worth emphasising that the thunderstorm also caused serious flooding on the southern and eastern side of Bowland.

A major flood occurred at the headwaters of various streams draining the Howgill Fells in the summer of 1982. The estimated return period was once in 70 years. There was massive hillslope erosion and opening up of grassed over gulleys with the creation of new erosion scars and a corresponding creation of new depositional features including debris cones and alluvial fans.[15] Thunderstorms causing flash floods do not happen only over remote upland valleys. When flash floods occur in towns their effects can be even more damaging with the loss of stock in shops and factories as well as damage in peoples' homes.

Most of the floods we have been considering in this chapter occurred in remote, thinly populated areas and affected only limited numbers of people. On the other hand the coasts of the north west are much more densely occupied, especially the Lancashire coast between Morecambe and Formby. In the next chapter we look at how flooding by the sea has affected these communities.

Notes

1   H. S. Cowper, *The Oldest Register Book of the Parish of Hawkshead in Lancashire 1568–1704* (Bemrose, 1897), pp. lx–lxi.

2   P. Carling & K. Beven, 'The hydrology, sedimentology and geomorphological implications of floods: an overview', in P. Carling & K. Beven (eds.), *Floods. Hydrological, Sedimentological and Geomorphological Implications* (Wiley, 1989), pp. 1–9.

3   G. Smith, 'Dreadful storm in Cumberland', *Gentleman's Magazine* 24 (1754), pp. 476–7; W. Hutchinson, *The History of the County of Cumberland and some places Adjacent,* (1794), vol I, p. 196; W. Gilpin, *Observations on the Mountains and Lakes of Cumberland and Westmorland* (1786) vol II, p. 36.

4   CRO (K), WQ/SR/215/23 1750, 220/11, 1750–1 213/19 1750.

5   Gilpin, *Observations,* vol II, pp. 4–6.

6   R. Doe, *Extreme Floods: a History in a Changing Climate* (Sutton, 2006), p. 132.

7   *Lancaster Guardian* 16 July 1870, 23 July 1870.

8   CRO (K), WPR 60.

9   P. Carling & M. S. Glaister, 'Reconstruction of a flood resulting from a moraine dam failure using geomorphological evidence and dam break modelling', in L. Mayer & D. Nash (eds.), *Catastrophic Floods* (Allen & Unwin, 1987), pp. 81–99.

10   *Westmorland Gazette* 5 Nov. 1927; *Penrith Observer* 1 Nov. 1927.

11   I. Tyler, *Greenside and the Mines of the Ullswater Valley* (Bluerock, 2001), pp. 73–76; Carling & Glaister, *Moraine Dam Failure*; S. Murphy, *Grey Gold. Men, Mining and Metallurgy at the Greenside Lead Mines in Cumbria, England 1825–1962* (Moiety, 1996), p. 312; *Penrith Observer* 25 Aug. 1931.

12   Doe, *Extreme Floods.*

13   *Westmorland Gazette* 4 July 1953, 11 July 1953.

14   E. Garnett, *The Wray Flood of 1967* (Centre for North-West Regional Studies, 2002).

15   A. D. M. Harvey, 'The river systems of North West England', in J. J. Johnston, *The Geomorphology of North West England* (Manchester University Press, 1985), pp. 122–42; A. D. M. Harvey, 'Geomorphic effects of a 100–year storm in the Howgill Fells, North West England', *Zeitschrift fur Geomorphologie* 30 (1986), pp. 71–91.

CHAPTER 8

Coastal Flooding

The coasts of the north west, with a few small exceptions like Humphrey Head on the north side of Morecambe Bay and St. Bees Head in west Cumbria, are low and made up of sediments such as sand and alluvium or unconsolidated deposits like glacial boulder clay. These can readily be eroded by wind and tide. For centuries coastal communities in this region have battled with the sea, sometimes extending the area of reclaimed land, sometimes seeing settlements washed away by storms or buried by blowing sand. Around Morecambe Bay, in the Fylde and between the Ribble and the Mersey there are many tales of lost communities. The medieval manor of Argarmeols in the parish of North Meols appears to have been overwhelmed by the sea at some time in the early fourteenth century by one of the great storms which heralded the start of the Little Ice Age.[1] The settlement of Ravensmeols, on the same stretch of coast, seems to have been buried by sand. In the Fylde tradition records the villages of Waddum Thorp near Fairhaven, south of Blackpool, and Singleton Thorpe near Rossal, which are popularly believed to have been swept away by the sea although historical evidence is lacking. More substantiated is Fordebottle in Furness which is thought to have been washed away by a high tide in December 1553.[2] Some of these disasters were, however, due to blowing sand and coastal erosion, rather than floods, which are our concern here.

Coastal and estuarine flooding usually occurs due to complex combinations of high tides, storm surges due to atmospheric pressure patterns, the action of storm waves and the backing up of rivers. Figure 44 shows the pattern of recorded sea floods over time. As with previous graphs and tables used in this book, it needs to be interpreted with care for the quality of the recording falls as you go back in time. For the seventeenth century only occasional severe events are recorded while for the twentieth century even minor overtopping of sea defences is mentioned. The graph does, however, suggest that some periods, for example the 1790s, the late nineteenth century and the 1920s, were relatively bad for coastal communities.

The worst sea flood on record, in terms of loss of life and the extent of the damage occurred on 18 and 19 December 1720. The sea broke through the dune barriers all along the coast between Formby and the

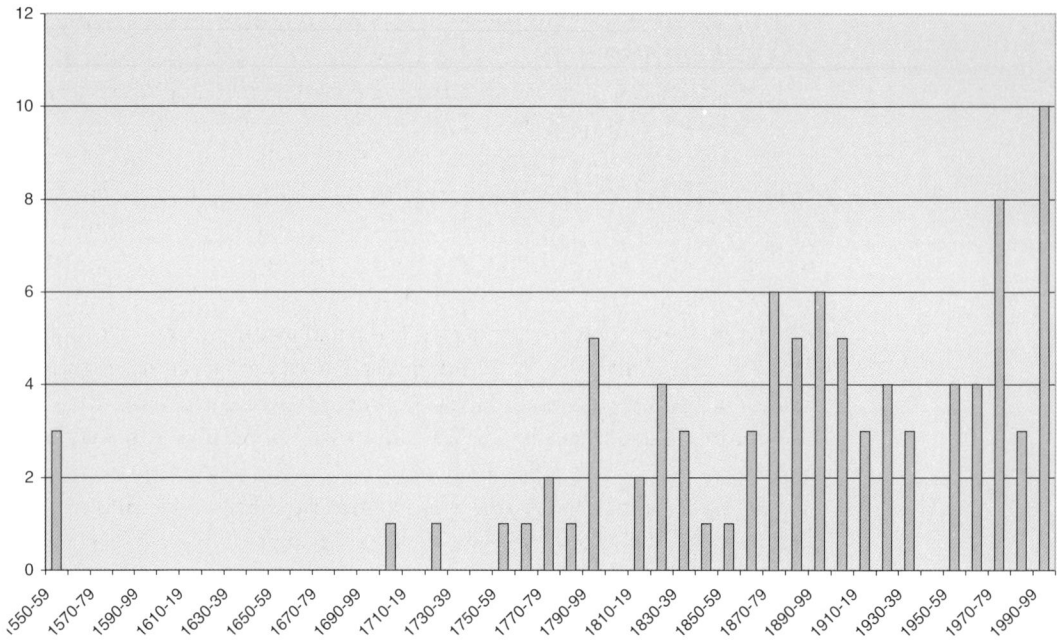

44: Known sea floods in the north west AD 1500–2000. Ian Whyte.

Ribble estuary as well as in the Fylde, the Lune estuary and around Morecambe Bay. Timothy Cragg's diary recorded that seven or eight people were drowned in the Cockerham–Pilling area while nine were killed in North Meols.[3] The total loss of life is likely to have been upwards of 20. Overall at least 157 houses were washed away and hundreds more damaged. Corn, hay, livestock, peat stacks, household goods, clothing and money were swept away with them, completely ruining many families. Extensive areas of arable land, sown for winter wheat, were damaged by salt water and the crops ruined. Uncounted livestock were drowned. In Pilling 15,000 acres of farmland were submerged and a further 5,000 in North Meols. A single farmer, Richard Jump from Alt Grange, had eight acres of wheat covered with sand and ruined, and 27 acres of ploughed land damaged by salt; his total damage bill was £185.10.0. The overall damage was estimated at more than £10,000 – a huge sum at this time and equivalent to some £1.2 million today. A number of bridges were also demolished.[4]

On 6 January 1791 Timothy Cragg described what was probably the worst sea flood since 1720. At Lancaster the tide covered the quayside and washed away many boats. At Pilling it breached a new dyke and a large number of sheep were drowned. In the autumn of 1795 an even worse flood broke the embankments at Cockerham, Pilling and Thurnam. Fields of wheat and potatoes were ruined, with water supplies as well as the soil contaminated with salt. At Pilling Hall cattle in the cowsheds were up to their bellies in water and at Lancaster the lower part of the town, along St. George's Quay, was flooded.[5]

The late eighteenth and early nineteenth centuries was a great age of reclamation from the sea in the north west, especially in the Ribble estuary where the Hesketh marshes were enclosed in a series of stages,[6] and around Morecambe Bay. Some areas of coastal salt marsh and peat moss were reclaimed under act of parliament, as in Cartmel and the Lyth Valley and Milnthorpe Marsh (Figure 45).[7] The reclaimed land was protected by embankments. Other areas were reclaimed by the initiative of individual landowners, such as James Wilkinson of Castlehead near Lindale. There were even more grandiose schemes to reclaim the whole of Morecambe Bay by constructing a railway embankment across it.[8] On 30 January 1877 Lancaster and Morecambe experienced another bad sea flood and the sea wall at Morecambe was partly washed away with low lying areas inland from it flooded. The town's new pier was also badly damaged.

On 23 March 1907 the bank protecting Milnthorpe Marsh was breached and the sea swept into College Green farm reaching almost to the level of the first floor where the farmer, 84 year old Mr. Lancaster, was bedridden (Figure 46). The dramatic events associated with this flood were used by Constance Holme in a fictionalised version for the climax of her novel *The Lonely Plough,* published in 1914. She had spoken to many of the people who had experienced the flood and her description vividly conveys the experiences of the local farming population:

> (He) saw ahead of him towering over the (sea bank) a white mountain of water as if the whole of the tidal wave had swerved and mounted its barrier. On a screamed prayer he turned and raced for his life with the monster behind him, and, as he reached the gate, a galloping horse

45: Part of the drainage system in the Lyth Valley reclaimed from peat moss in the early nineteenth century. Photograph Ian Whyte.

46: Some of the low lying land near Whitbarrow flooded in 1927. Photograph Ian Whyte.

and rocking trap burst past him into the yard, a flood also at its heels. The water poured after them up the slope, and above the shriek of the wind they heard the roar of the full tide as it swung on and past to the top of the bay.

In the kitchen, the frightened women and the roused hands were busy moving food and valuables as the sea came in at the door, until presently it was standing two feet deep on the flags ….. 'The banks are giving on all sides.' Michael said as the household crowded round him 'We were near caught time and again as we came along … The marsh road's gone, ay an' the main road an' all, I doubt.'

An attempt was made to reach the outlying farm of Ninekyrkes:

They formed a human chain and groped, with the big sticks scouting before them, in imminent danger all the time, and more than once utterly bewildered and all but lost. Wading often to their waists, trapped by deep holes, by wire fencing wrenched into sunk snares for their stumbling feet … buffeted by the wild gusts and clouds of spray beating in through the mighty breaks in the bank, they yet held on … and knew that the floodgates at Ninekyrkes must be smashed.

After the flood had subsided the full extent of the damage was revealed:

On the sea-roads the water rose level with the hedges all day, and, when it left, the scars of the land crept shudderingly into sight. Great holes five and six feet deep where had been metalled surface, uprooted fences and railings twisted like cord; and everywhere dead things, rabbits, horses, poultry – and always sheep. The peaceful, cared-for country lay broken and horribly disfigured, as if by the riving hands of a maddened giant.[9]

If the sea flood of 1720 was the worst in the north west to affect a rural area then the one which hit Fleetwood on 28 October 1927 was probably the worst urban disaster. Fleetwood had been built from 1836 as a new town and port among the sand dunes at the mouth of the Wyre. Rudimentary coastal defences had been constructed but these were washed away by a flood in 1863. Stronger embankments were built at the end of the nineteenth century but the disaster of 1927 was blamed by some on the removal of gravel, shingle and sand by the local authorities in Blackpool to strengthen their own sea defences. On the night of 28 October, with a storm and an exceptionally high tide, the sea defences were breached just north of Rossal School and the floodwater spread out across the peninsula on which the town was built with water up to 12–feet deep in some streets. Six people were drowned, most of them at a caravan site where the caravans were completely washed away by the tide. In the lowest lying areas some families had to climb on to the roofs of their houses for safety and wait to be rescued.[10] The damage and danger was increased by logs from a flooded sawmill being washed around the streets. Around 1,200 homes were seriously damaged with water up to 10 or 12 feet deep. Six hundred families were marooned in the upper floors of their houses for several days and keeping them supplied with food and fuel occupied a large number of people in small boats. Of a total population of 22,000 in the town c. 10,000 were affected badly. The flooding was so severe that no tram, train or bus could get into Fleetwood for several days. The first relief train from Wigan reached Fleetwood over submerged tracks at a walking pace to allow debris to be cleared off the line. There was water in the streets for a week but in the outlying districts it took a month to remove it all.[11]

Whilst the worst effects of this storm were undoubtedly at Fleetwood the whole Lancashire coast was affected. The sea breached a new £60,000 earth and turf embankment at Meathop on Morecambe Bay which was 50–feet wide at the base and five feet thick at the top. It had allowed around 7,000 acres to be reclaimed but the sea flooded in once more with heavy loss of sheep and cattle, leaving a foot of mud in flooded homes when the tide retreated. The Foulshaw and Sampool embankments were also breached. Some £8,000 worth of damage was caused to Blackpool promenade and there was extensive flooding in Preston and Penwortham. At Lytham over 50 houses were badly flooded so that the local authorities there started their own flood relief scheme. The Ribble breached the embankments protecting Hesketh and Longton

marshes drowning more livestock. The Wyre burst its banks for the third
time in five weeks around Garstang and St. Michaels, cutting several
farmhouses off.

One of the most remarkable transformations in the landscape of the
north west was the reclamation of the Lancashire mosslands. Apart from
Chat Moss and Trafford Moss and other smaller mosses near Manchester,
there was a continuous belt from near Liverpool to the Ribble in the
middle of which was the lake of Martin Mere, shown on county maps
from the late sixteenth century survey by Christopher Saxton onwards.
At its greatest extent Martin Mere may have been five miles long and two
wide, draining not directly to the sea but eastwards into the River
Douglas. In the Fylde, Pilling, Rawcliffe and Stalmine mosses covered at
least 20,000 acres.

In 1692 Thomas Fleetwood of Bank Hall obtained an act of Parliament
permitting him to commence drainage operations by cutting a mile and a
half long channel to the sea at Crossens, employing up to 2,000
workmen, a considerable engineering feat in its day. By 1700 most of the
lake had been drained. Progressive silting up of the channel in the early
eighteenth century impaired drainage and winter flooding took longer
and longer to clear. In 1755, however, a very high tide washed down the
flood gates and Martin Mere became a lake again. In 1781 Thomas
Eccleston of Scarisbrick Hall employed an engineer named Gilbert, who
had worked for the Duke of Bridgewater at Worsley, to make another
effort. He cleared and deepened the original channel installing three sets
of flood gates as backup. This was a sensible idea for in 1813 an unusually
high tide swept away the outer gates but the others held and there was no
serious flooding.[12] Large-scale drainage in the northern part of the Fylde
started later, in 1830, when Wilson France of Rawcliffe Hall began to
reclaim the mosses on his estate by cutting a dyke six miles long to the
sea.

Rockcliffe marshes on the Solway estuary experienced a disaster in
February 1967 when a 150–year old flood bank burst. Hundreds of sheep
were drowned, the worst livestock losses on record in this area. The bank
was owned by the Castletown estates which were responsible for its
maintenance and repair. A previous flood 26 years before had resulted in
insurance payments which were so high that companies had refused to
renew the policies for farms on the marsh without unaffordably high
premiums so that most of the farmers were uninsured.[13]

With rising sea levels and more surges in prospect the work of
protecting the coasts of the north west continues. In 2005 work started
on a £26 million improvement to the coastal defences at Cleveleys which
is designed to protect the resort from the sea as well as enhancing its
appearance. It is designed to protect Cleveleys from flooding by the sea,
safeguarding over 700 homes and 200 commercial properties, and should

47: Cleveleys Seawall, April 2008. Reproduced with thanks to Wyre Borough Council and Birse Coastal.

48: Cleveleys Seawall, April 2008. Reproduced with thanks to Wyre Borough Council and Birse Coastal.

be completed by 2010 (Figures 47 & 48). Other schemes are in progress or are planned but the level of investment needed is likely to rise substantially in the next few years. In November 2007 on the east coast of England a combination of pressure, wind and tide conditions produced sea levels which came within a few inches of overtopping the defences. A similar set of circumstances could occur in the north west, making emergency planning and aid for flood victims an important concern for local authorities and – as we will see in the next chapter – for voluntary organisations.

Notes

[1]   C. W. Farrer, *A History of the Parish of North Meols* (Henry Young and Son, 1903), pp. 97–8.

[2]   W. Ashton, *The Battle of Land and Sea on the Lancashire, Cheshire and North Wales Coasts* (Heywood, 1909).

[3]   LRO, DDX 760/1.

[4]   J. Beck, 'The church brief for the inundation of the Lancashire coast in 1720', *Transactions, Historic Society of Lancashire and Cheshire* 105 (1953), pp. 91–105.

[5]   LRO, DDX 760/1.

[6]   A. J. L. Winchester & A. G. Crosby, *England's Landscape. The North West* (English Heritage, 2006), pp. 102–3.

[7]   I. D. Whyte, *Transforming Fell and Valley. Parliamentary Enclosure and the Landscape of North West England* (Centre for North-West Regional Studies, 2003).

[8]   W. Rollinson, 'Schemes for the reclamation of land from the sea in North Lancashire during the eighteenth and nineteenth centuries', *Transactions of the Historic Society of Lancashire and Cheshire* 115 (1963), pp. 133–45.

[9]   C. Holme, *The Lonely Plough* (Oxford University Press, 1931 ed.), pp. 295–6, 301.

[10]   *Lancashire Evening Post*, 2 Nov. 1927.

[11]   Fleetwood Public Library, Flood Files of press cuttings.

[12]   R. Millward, *Lancashire* (Hodder, 1955), pp. 51–2; W. G. Hale & A. Coney, *Martin Mere. Lancashire's Lost Lake* (Liverpool University Press, 2005), pp. 125–50.

[13]   *Cumberland Journal* 3 March 1967.

CHAPTER 9

Flood Relief

Today, with the insurance of property and house contents widespread (though far from universal), and with local authorities and government agencies geared up to coping with disasters, there is less need for privately organised relief for flood victims than in the past. However, voluntary organisations still play a very important role in coping with the immediate turmoil created by flooding and its longer term impacts. In the past local voluntary aid was more urgently needed because of a lack of insurance and the knowledge that neither national nor local state agencies would be able to do much to help.

We know little of the extent of assistance provided for flood victims by the pre-Reformation church whether from monastic houses or at parish level. In the seventeenth and eighteenth centuries, however, a modest amount of direct aid might be provided by parish churchwardens, as at Arthuret near Carlisle in 1687 and 1701.[1] The Quarter Sessions were more concerned with the repair and maintenance of flood-damaged bridges but they might also consider petitions for aid from those affected by floods. In 1704 Henry Wilkinson pleaded to the Carlisle Quarter Sessions that his cottage and field had been damaged by a flood and his cow drowned. In 1733 John Pears claimed that his wheat crop had been swept away.[2] Borough authorities might also provide help, as in 1809 when the Chamberlain of Carlisle granted £2.2.9. to the victims of a recent flood.[3] After the great sea flood of 1720 the communities affected petitioned the Crown for letters authorising them to make voluntary collections in churches and from door to door in Lancashire and other counties.[4]

The role of private aid remained considerable in the eighteenth and nineteenth centuries. In 1792, after a flood on the River Eden, the *Cumberland Pacquet* was of the opinion that 'Disastrous events like this call loudly on the compassion of the landlords'.[5] In February 1822, when Carlisle suffered severe flooding (Chapter 6), it was not the council who organised a relief fund but 'some young ladies of Carlisle' who raised a subscription and organised the distribution of coal and food to affected people.[6] Indeed there seems to have been reluctance on the part of urban corporations and local authorities to accept any responsibility for flooding at all or to provide any aid and compensation for victims.

As the nineteenth century progressed, however, it became more common to establish local relief funds to cope with flood disasters. After floods in Carlisle in 1822 and 1856 funds were created to help the inhabitants of Rickersgate who had suffered the worst damage in the city. In 1856 individuals received up to five shillings each or blankets and coal.[7] Nevertheless many nineteenth-century floods are recorded where the families affected seem to have been left to cope as best they could or to depend on assistance from family and friends,

Even at the end of the nineteenth century, as is made clear by the records of the Garsdale Inundation Fund (Chapter 7), flood relief was still ad hoc, local and small in scale (Figure 49). The construction of a range of institutional buildings in nineteenth-century towns such as schools, chapels and community halls made it easier for urban authorities to provide temporary accommodation for those evacuated from their homes. In Carlisle in 1925, when a number of people were made homeless after a flood, two schools were opened to house them for a few days while soup kitchens were also organised and teams of volunteers recruited to take food round to people who were still marooned in their homes.[8]

The flooding of Fleetwood in October 1927 (Chapter 8) was one of the most damaging disasters in the north west in modern times and prompted a nationwide appeal for aid. The plight of thousands of the inhabitants was unprecedented in its severity. Two houses even collapsed as a result of the floods and instances of looting were reported. A relief fund was opened which attracted donations from the Prince of Wales, newspapers and other Lancashire towns. Over £107,000 was eventually

49: Farm in Garsdale which was damaged by the 1889 flood. Photograph Ian Whyte.

raised.[9] A minute book for the flood relief committee which administered the fund has survived.[10] Membership of the committee included not only local councillors but representatives from various churches, the Salvation Army, local industry and Superintendant Crapper of the police. While much of the town was still under water the committee arranged a system of distributing coupons which flood victims could exchange for food and coal to the value of five shillings per adult per week for food or two hundredweights of coal. The vouchers could also be exchanged for household goods. Staff at local schools were asked to undertake the distribution of the coupons. On 4 April it was reported that the director of Bryn Hall colliery, Wigan, had offered 100 tons of free coal. Temporary accommodation was arranged for those caravan dwellers who had lost their homes entirely. People from the furnishing and carpet trades were recruited to act as voluntary valuers for assessing the extent of damage to peoples' possessions. The city of Leeds arranged a convoy of lorries which left on 16 November bringing donations of furniture and other necessities. Householders were compensated for the loss of their possessions initially at a rate of 13.4d. in the pound. Tradesmen, farmers, allotment holders, pig and poultry keepers received 10 shillings in the pound. Householders later received another 6.8d. in the pound so that the full costs of damage were met. Although the minute book rarely goes into personal details some entries highlight the individual losses which were covered; grants of new textbooks to evening school students and new equipment for midwives. Payments to householders included an allowance for doctors' bills and help with overdue interest payments on mortgages. Whether or not the local council bore any responsibility for the flood through neglect of the town's sea defences the relief operations appear to have been large-scale and efficient. Because of the national scale of the appeal the level of donations seems to have been equal to the scale of the damage. The War Office put a large barracks and military hospital at the disposal of the council to house people who had been evacuated from their homes.

The hero of Fleetwood's floods was undoubtedly Superintendant Crapper who co-ordinated all the measures to keep people who were trapped in the upper floors of their homes supplied with food, drink and fuel. His wife, meantime, organised the collection and distribution of clothes for the flood victims. Tremendous spirit was shown by the local people and the police station was besieged by volunteers offering help.[11] When a shortage of small boats to navigate the streets became evident Crapper did some inspired lateral thinking and had 40 small rowing boats brought from a lake in Stanley Park, Blackpool. They were taken by lorry to the edge of the floods. Volunteers, including boys from Rossal School, then rowed and dragged the boats through flooded fields and streets, negotiating hazards like submerged spiked railings, to the centre of Fleetwood. It must have been an exciting adventure for the schoolboys.

A vivid picture of the damage caused by flooding to ordinary households in the past comes from the records of the Walton le Dale Flood Relief Fund, established after a severe flood in September 1946 by the River Ribble and its tributaries.[12] Just over £2,035 was collected. The 205 people who submitted claims to the fund were required to fill in forms with details of their losses and damage. These forms provide a fascinating insight into the kind of flood damage that occurred just before the start of post-war prosperity and consumerism. The depth of water in various houses ranged from two to four feet or five feet in back yards. Less than one in five of the claimants had their house contents insured. Many of those who were insured were with a free scheme organised by *John Bull* magazine where there were no premiums but very small payouts. Another quarter were insured either for very little or for fire but not flood. The bulk of the claims were simply for ruined or damaged floor coverings, especially linoleum and underfelt. In a period of post-war austerity damage to carpets and furniture was mentioned in terms of costs of drying and repair rather than replacement. Apart from furniture the items most frequently mentioned as damaged were pianos – which could hardly be moved upstairs quickly when a flood threatened. Some claims mentioned damage to wireless sets or gramophones, only one or two electric washing machines or cars.

Most of the houses were two-bedroomed terraced properties (Figure 50). Some people were clearly unable to afford to replace their losses. One of the saddest cases was a man whose only claimed loss, apart from floor coverings, was three concertinas (replacement value £40–60 each) and two 'American organs' with a note asking that the assessor should take into account that these instruments represented his livelihood. There was a sympathetic letter accompanying the form from a firm in London confirming that the instrument that had been sent to them was not repairable. In one house a daughter's essays, notes and papers from university had been destroyed. There were many claims for damage to wallpaper, plastering and skirting boards – though a suspiciously large number of people claimed to have only just completed redecorating when the flood occurred. The farmer at Flatts House Farm claimed £400 for the loss of his market garden crops, vegetables and flowers emphasising Walton le Dale's local speciality. He was awarded only £34 for household contents.

Many boundary fences and garden walls had been demolished and required rebuilding. In a number of cases wooden floors had to be replaced. One house had its foundations partly undermined with earth being washed away at the side of the house to a depth of five feet, requiring eight tons of material to fill the hole. In other cases foundations had started to subside.

50: Some of the houses in Walton le Dale, a stone's throw from the Ribble, which were flooded in 1946. Photograph Ian Whyte.

One or two people in their claims mentioned health problems and nervous stress resulting from the flood while two claimants recorded their preference that the money from the fund be used in preventing further flooding. One man blamed the flooding from the River Darwen on the raising of a weir on two occasions by Messrs. Lees, with the permission of the council (Figure 51).

51: The River Darwen at Walton le Dale. Photograph Ian Whyte.

Receipt books recording 450 donations to the fund have survived. Many were small sums from individuals with local councillors prominent among the earlier gifts. The rest came mainly from local businesses in Walton, Bamber Bridge and Preston (though a good deal of this money was donated by employees rather than their employers). Some of the money was raised from local events like jumble sales, dances and boxing matches. Occasional donations came from further afield, including firms in Birmingham, Bristol, Nottingham and London – presumably ones with some local links.

Following this flood, and an earlier one in December 1936, Walton le Dale acquired the label of 'flood black spot' in the local press. When the Lynmouth flood disaster occurred in 1952 a local relief fund was opened. As a member of the council commented 'We in Walton know what flooding is'. A number of local firms whose premises had been damaged in 1946 made donations.[13] This emphasises the point that the social impact of foods may not be completely disastrous if they help to reinforce community solidarity.[14] Walton's reputation seemed to be confirmed when a third flood occurred in January 1954. Another flood relief fund was set up although the scale of the damage was considerably less than in 1946. Only 40 claims were made, most of them reasonable ones according to the valuers. The scale of response was also smaller despite a flag day to raise money and only £352.19.0. was collected.

In the 1950s and 1960s the response to helping people suffering from flood damage was still essentially ad hoc and local though the 'Lynmouth effect' must have increased public concern and awareness. In 1947 in Appleby the British Legion gave 11cwt of coal to each ex-serviceman who was flooded. As a result of the floods of 1954 and 1964 a more permanent scheme known as the Mayor's Fund was established, the first official response in Appleby to the flood hazard. In December 1954 an appeal fund was opened in Kendal for local flood victims.[15]

Damage sustained in the Troutbeck flood of 1953 was met by a relief fund organised by the Lakes District Council, though for the Wray flood of 1967 a national appeal was launched. The money which these appeals brought in was still inadequate to cover the damage at a time when many people did not have their homes or contents insured. After the severe floods of Borrowdale in 1966 an appeal raised £2,828 but the total damage was estimated at around £200,000 (Figure 52).

On 18 July 1964, a day of severe thunderstorms across many parts of Lancashire, a flood affected the east Lancashire town of Oswaldtwistle. A relief fund was opened which eventually raised £442.17.6. A surviving receipt book shows that around three-quarters of the money came from local businesses, clubs, societies and churches. The rest came from individual donations, collection boxes, and charity events including cricket matches and a mannequin parade. Payments were made to 37

52: Aftermath of a flood in Borrowdale, 1966. Cumbria County Council Library Services, Carlisle.

addresses; the names of some of which, including Waterside Bungalow and Brookside Cottages should surely have warned the occupants. The claim forms show that only around a quarter of the households were insured. No family received more than £15, and most got under £7. The expense of replacing carpets had risen markedly since the Walton le Dale flood of 1946 with fitted carpets now being much more common and costing £75–100 to replace. Expensive electrical appliances like washing machines were also more common.[16]

The spread of insurance cover for houses and their contents since the 1960s has reduced the need for private flood relief organisations and may have made some aspects of flooding easier to bear but it can still be a traumatic experience. In the next chapter we look at how successful north west communities have been at protecting themselves from floods in the past. With recent floods in the Midlands and eastern England in mind we also consider the state of flood defences today.

Notes

[1]   CRO (C), PR 18/18.

[2]   CRO (C), QS 11/1/174/12, 11/1/171/12.

[3]   *Carlisle Patriot* 9 Feb. 1822.

[4]   J. Beck, 'The church brief for the inundation of the Lancashire coast in 1720', *Transactions, Historic Society of Lancashire and Cheshire* 105 (1953), pp. 91–105.

[5]   *Carlisle Patriot* 4 Sept. 1792.

6   *Carlisle Patriot* 9 Feb. 1822

7   K. Smith & G. A. Tobin, *Human Adjustment to the Flood Hazard* (Longman, 1979).

8   *Carlisle Journal* 2 Jan. 1925.

9   *The Times* 1 Nov. 1927, 8, 10, 18, 20, 22 Nov. 1927; Fleetwood Library files on the 1927 flood.

10   LRO, MBF 11/11.

11   *Lancashire Evening Post* 29 Oct. 1927, 1 Nov. 1927.

12   LRO, UDWd 37/3.

13   LRO, UDWd 37/3.

14   F. Furedi, 'The changing meaning of disaster', *Area* 39 (4) 2007, pp. 482–89.

15   *Westmorland Gazette* 11 Dec. 1954.

16   LRO, UDOs 18/16.

CHAPTER 10

Controlling Flooding: past, present and future

Although the damage caused by flooding seems to have risen steadily from the nineteenth century with the spread of floodplain developments, flood protection measures are in many areas a comparatively new phenomenon.

Early forms of flood protection were often unobtrusive such as the careful siting of settlements and farmsteads out of reach of potentially high river levels and building hump-backed bridges with high stone arches which would withstand the worst inundations. Monastic landowners are said to have protected some of the low-lying land between Brothers Water and Ullswater in the Lake District with embankments. On the River Alt there had been embanking against floodwater as early as the thirteenth century on Whalley Abbey's land at Alt Grange. Before the later eighteenth century there was more concern to protect coastal areas from flooding by the sea, or by streams whose lower courses might be backed up by high tides. A study of the River Alt near Liverpool.[1] shows that in 1589 a commission issued by the Duchy of Lancaster to enquire into the danger of the sea breaking the embankments at Alt Mouth recommended that they should be strengthened with a ten-foot high wall over half a mile long at a cost of £660. There is no indication that this wall was actually built but the southern hundreds of Lancashire were certainly forced to contribute towards it. The medieval institution of the Commissioners of Sewers were sometimes active in flood defence. A commission for the Alt is known from 1608 when its main activities were improving embankments and scouring the channel of the river to improve the streamflow. Shrinkage of the peat surface on reclaimed land was encouraging flooding. Wet weather and disruption during the Civil Wars also seems to have caused problems as in 1655, 1657 and 1659 local people petitioned Cromwell for relief from taxes because of the floods resulting from breaches in various embankments. New commissions were issued in 1660, 1677, 1695 and 1695 to tackle the problem.[2]

Elsewhere some flood protection measures came early where there was a specific flooding problem. At Preston the Ribble Bridge at Walton le

Dale seems to have been in danger of being abandoned by the river which was attempting to change its course to the south, leading to efforts to stabilise the south bank and reduce erosion. Sometime in the sixteenth century a protective bank or cop was built here to stabilise the river and reduce the flood risk. In 1634 local magistrates agreed to spend money on repairing it. The bank gave its name to the extensive level alluvial area between the confluence of the Ribble and the River Darwen, rich land, ideal for market gardening in later centuries, but which was always vulnerable to flooding (Figure 53). In 1669 the bank was described as having been damaged by recent floods and it was breached by another one in 1679. The bank continued to be maintained, with major rebuilding in 1822 and its modern counterpart is the Environment Agency's wall of 2002 (Figure 54). It is claimed that the central part of the area between the Ribble and the Darwen was deliberately not built up to allow it to absorb flood water and reduce damage elsewhere.[3] If so, this is an early example of the modern practice of building flood storage. In the early eighteenth century there was a line of stone posts and rails from the Walton Bridge over the Ribble along the downstream end of the road leading across the cop (close to or a little to the south of the modern Victoria Road). The aim was to provide something solid for people to grab hold of when the road was submerged by floodwater, to reduce the risk of their being washed away (Figure 55).[4]

During the eighteenth and nineteenth centuries flood defences in urban areas were small-scale and piecemeal, often put up privately to protect individual factories and mills. Some measures to reduce flood risk

53: Flood bank at the Capitol Centre, Walton le Dale. Photograph Ian Whyte.

54: The Ribble Bridge at Walton le Dale with 2002 flood wall. Photograph Ian Whyte.

55: Early eighteenth century print of the Ribble Bridge, Walton le Dale, with line of posts to the right. Reproduced by permission of Lancashire County Library and Information Service. http://lanternimages.lancashire.gov.uk/No. 773.

were taken quite early. In Carlisle there were small embankments in Rickersgate by 1851 which are reported to have saved the area from flooding in 1851 and 1856.[5]

In the second half of the nineteenth century urban authorities were more active. Their policy was that floods should be controlled rather than that further development on the flood plains should be prevented. In Carlisle by the end of nineteenth century it was well known that flooding was a recurrent problem in the city every 10–15 years or so. In the early twentieth century several large embankments were raised around Bitts

Park and the Sauceries (Figure 56). After a flood in 1931 the channel was widened at Petteril Bridge and parts of the river's course straightened. A fragmented approach, responding only to the most recent events, was adopted until the flood of 1968. Work on the Eden between 1947 and 1952 cost £200,000, and mainly involved deepening of the channel. In 1954 engineers of the river board reported that the channel of the Eden had been lowered over three feet and its flow capacity raised by 25 per cent. The flood of 1968 led to widespread criticism of the inadequate level of flood protection in Carlisle. Following this there were structural measures to alleviate floods and better forecasting and warning of their occurrence. Existing measures were extended into a comprehensive embankment scheme which was intended to be proof against a one in 100–year flood (including the flood of 1968). The work was completed by 1971. In practice the defences were not as effective as had been hoped. It was suggested that there was likely to be a problem within the embanked areas due to water backing up in the drains and that there was a danger of further building going ahead within the protected area. The floods of January 2005 showed that these defences were certainly not capable of dealing with an event with a return period of once in a century or more.

In Appleby no major flood alleviation scheme was in place as late as 1979. After flooding in 1928 and 1929 various minor structural schemes were proposed but nothing was done. Following the 1968 disaster there was liaison between local authorities and the Cumberland River Authority. A scheme to deepen the Eden from the main bridge downstream to Holme Farm with a system of graded banks along the Sands was proposed at an estimated cost in 1975 of £170,000, but was not adopted because of the expense. The response after the 1968 flood was

56: Flood embankment at Eden Bridge, Carlisle. Photograph Ian Whyte.

little more than a reconstruction of the damaged properties, repair of banks along The Sands and the rebuilding of the Jubilee Bridge which had been washed away. In 1993 a £1m flood relief scheme was rejected because of fear of its effects on the town's historic character. The National Rivers Authority sought permission to carry out work at King George's Field, Chapel Street, the Cloisters and St Lawrence's church with new banks and flood gates, and in 1995 a £750,000 flood alleviation scheme was finally opened which was designed to protect Chapel, Bridge and Holme Streets (Figure 57).[6]

In Kendal it was noted on 12 February 1831 that if it had not been for the building of the Mill Bridge and improvements to the banks in recent years the flood at that date would have been even worse.[7] The history of flooding in Kendal from the eighteenth century until the 1930s was one of growing interference with, and impeding of, the course of the River Kent, increasing the risk of flooding for a given volume of discharge. Since the 1930s there have been increasing efforts to widen and clear the river channel to reduce the flood risk, together with growing protection of the banks. The construction of weirs to supply a head of water to grain, fulling and other kinds of mills, as well as bridge construction, may have had some effect in increasing the flood risk before the mid-eighteenth century. It was, however, the building of a series of new large mills along the banks of the Kent in the first two decades of the nineteenth century which started to impede the channel significantly. The building and rebuilding of bridges was also a feature from the 1740s to the 1820s. The creation of streets of mill workers' houses may also have contributed to the flood risk by increasing the local rate of runoff. All these trends are likely to have added to the volume of floodwater and

57: Part of Appleby's most recent flood barrier Photograph Ian Whyte.

increased its obstruction. More widely within the catchment, the drainage of Kentmere Tarn between 1828 and 1836 was thought to have increased the susceptibility of the Kent to sudden floods. Undersoil drainage from the 1840s onwards may have done the same. In the twentieth century the improved drainage of roads was also considered thought to have increased the flood rate.

In Kendal (Figure 58) larger-scale measures to reduce the risk of flooding were first made in 1933 when the weir at Low Mills was reduced in height.[8] This is claimed to have saved Kendal from serious damage by flooding in 1938.[9] In the wake of five floods in rapid succession during the autumn and winter of 1954 more serious measures were taken. Between the Nether and Miller bridges *c.* 5,000 cubic yards of gravel were removed from the bed of the river,[10] lowering the bed by at least 18 inches and in some places by up to five feet. In 1957 the weir at Stramongate Bridge was removed and replaced by a new one two feet lower. These measures were designed to prevent a volume of water equal to the worst of the 1954 floods from overflowing the river banks, though this was less than the worst recorded flood of 1898. Work on a further flood alleviation scheme began in 1972 at a cost of around £1 million. The work involved improvements on five km of river from Mintsfeet upstream of Kendal to Watercrook below it. The river channel was widened and deepened with some 240,000 cubic metres of material being removed. The existing river walls were strengthened and a further 1,700m of new walls built. To prevent the river bed from drying out at periods of low flow a series of small weirs was constructed across the stream (Figure 59). An

58: Aerial view of Kendal and the River Kent emphasising the town's vulnerability to flooding. © Environment Agency copyright 2008. All rights reserved.

59: The modern
channel of the Kent
at Kendal with weirs
to prevent drying out
in periods of low
water. Photograph
Ian Whyte.

embankment was raised to protect the Mintsfeet industrial estate to the
north of the town and a large lagoon constructed there as a gravel trap
(Figure 60). This scheme was opened in 1978. In 1985 a flood comparable
to the one of 1954 passed through Kendal but was contained by the new
defences.

Some rivers, such as the Alt near Liverpool, are already highly
engineered with extensive flood defence systems, to reduce the risk to
valuable agricultural land as well as urban areas. The defences on the Alt

60: Aerial view of
lagoon for
intercepting gravel at
Mintsfeet. ©
Environment
Agency copyright
2008. All rights
reserved.

61: The channel of the beck in Great Langdale is completely artificial, designed to move floodwater downstream as quickly as possible. Photograph Ian Whyte.

include pumping stations at the mouth of the river to prevent tidal flooding, several other pumping stations and embankments. Maintaining these costs the Environment Agency over £500,000 a year but a recent report suggests that further investment is needed.[11]

As we have seen with Kendal, one way to reduce the incidence of flooding is to alter river channels and make these more efficient, by deepening, widening, straightening and the removal of obstructions so that they can hold more water and move it through the system faster. Channel improvements and embankments remain a popular form of protection. A good example of a rural stream with a largely artificial channel is the beck that drains Great Langdale in the Lake District. With its steep valley head and high surrounding fells this valley is particularly prone to flooding and the wide, straighter, smoother channel reduces this risk substantially. Improving drains may not be a very gripping topic; raising water rates to pay for them is even less popular. For a century or more governments and local authorities have relied on the legacy of their Victorian predecessors to deal with the problems.

The normal approach to flood protection in the past has focused on 'hard engineering' solutions; building embankments, modifying channels and pumping drainage water out of reclaimed areas. Hard engineering solutions, however, address the symptoms rather than their causes. Sustainable drainage systems (SUDS) provided a more sustainable solution to the disposal of groundwater than culverts and drains. Sustainable drainage systems include, for example, permeable pavements which allow water to soak into the subsoil, or collecting water in basins or 'swales' to disperse into subsoil or wetlands. The aim is to retain water for as long as possible by slowing its dispersal, maximising the amount reaching aquifers. There is an increasing focus on the alternative approach of 'managed re-alignment', providing more space within river basins for storing floodwater. Washlands are areas adjoining streams which are allowed to flood in a controlled way. They are normally protected by flood banks and only fill up when the banks have been overtopped. On the River Wyre, after a major flood in 1980 affected 400 houses, a meander on a tributary, the River Brock, was cut off to speed up the flow of water. Embankments over three metres high were built some distance back from the river. An area of the floodplain east of the village of St. Michaels on Wyre was converted into a flood reservoir, farmland into which excess water could be diverted, with another scheme six km upstream at Garstang. The scope for creating flood reservoirs is, however, often limited by demands for agriculture and building. Farmers receive compensation when their land is flooded in this way but perhaps payments should be regularised so that stored floodwater could become a kind of crop.

Current work by the Environment Agency in developing flood risk management strategies for north west catchments aims to provide a framework for the next 100 years.[12] Environment Agency Catchment Management Plans can be downloaded from their website. The purpose of Catchment Flood Management Plans is to:

- Understand the factors that contribute to flood risk within a catchment such as how the land is used
- Recommend the best ways of managing the risk of flooding within the catchment over the next 50–100 years
- Catchment Flood Management Plans help to identify priorities for flood risk and the environment and so provide best value for money for the investment of public money.

When floods do occur, despite all the measures to prevent them, it is essential that early warnings are given and that people living in threatened areas know what to do and how to protect themselves and their property. The Environment Agency website lists flood warnings, updated every 15 minutes, for particular rivers but people need to be aware that the service is there. The flooding of Carlisle in 2005 emphasised not only the limits of the existing flood defences but also the widespread lack of awareness about the extent of the flood hazard. Studies of Carlisle and its surrounding villages since then have indicated that only a small proportion of the population affected had been flooded before and that householders consistently underestimated the risk of flooding. Even after the disaster two-thirds of the people in nearby villages such as Warwick on Eden who suffered damage thought it unlikely that another flood on this scale would ever happen again. None of the people in the village who had been affected by flooding were uninsured but many of them found it difficult to get re-insured after the flood and many were forced to pay higher premiums. The owner of the pub in the village, the Stag Inn, was unable to get his property insured at all. The cost of renovating and repairing existing houses ranged from £10,000–£140,000 each. But flood warnings are of little value unless everyone concerned – individual householders and local organisations alike – respond effectively when flood warnings are given. The villagers at Warwick on Eden were also targeted to make them more aware of the early warning systems which existed and of the measures they could take themselves to reduce the damage caused by floods.

Flood defences, even for quite limited areas, require multi-million pound budgets and take many years to plan, finance and construct.[13] The economic damage caused by flooding is receiving increasing attention in the media and from central and local government. There is a widespread belief that the incidence of flooding in Britain is likely to rise during the next few decades as a result of climate change and the growing development of floodplains. As it is impossible to defend everyone against all flood risks the need to think about floods in wider terms has become evident involving concepts such as improving warning systems and people's general awareness of flood hazards.

In recent years there have been significant changes in approach to managing flood risk in England and Wales. The old approach was predominantly urban in focus. In 2004 DEFRA signalled a new strategy 'Making space for water', a more holistic approach to flood risk management uses a range of approaches. It considers risk from all sources of flooding including floods due to inadequate drains and sewers. It sees a range of benefits from flood protection, for example improving conditions for wildlife. It uses a wide range of risk management options.

Nevertheless, there have been failings due to the complexity of the system that has been developed for flood protection in England and Wales. No single body is solely responsible for flood defence in England. The Environment Agency exercises an overall supervision over all matters relating to flood defence as well as having operational powers to carry out works on designated main rivers and on sea/tidal defences. In some parts of the country (but not the north west) there are Internal Drainage Boards which carry out works on designated watercourses in their area. OFWAT and water utilities are responsible for managing the risks associated with sewer floods. The highway authority has responsibility for culverts beneath and drainage from the highway. In the past there has often been relatively little dialogue between these various authorities. Also some defences are still privately owned. The Environment Agency is now taking on new responsibility for coastal flooding. Too many authorities are currently involved for coherent, consistent planning. What is needed is a single body to take overall responsibility for drainage and flood defence. Around 10 per cent of all houses in England lie within the once in a hundred year flood limit. Half the housing built in England since World War Two has been on floodplain land. A third of the areas earmarked for future housing are also on floodplains. In recent years the north west's record has been better than for some other parts of the country; around 8–9 per cent of new houses in the region have been built in flood-risk areas against 20–28 per cent in London; but there is no room for complacency. Hospitals and schools are still built on floodplains because they provide large areas of level land. This renders vulnerable to flooding precisely the institutions which are needed to help communities during potential disasters. Floodplain developments can have the effect of pushing floodwater into areas never previously affected by flooding.

Flood defence in recent years has been a failure of planning policies and inadequate precautionary measures. The National Audit Office reported in July 2007 that only 57 per cent of Britain's flood defence systems were in good condition and only 46 per cent of those protecting towns. Water companies are unwilling to build new storm water drains because they are not allowed to pass on the cost to the consumer. In England and Wales there has been a lack of integration between flood defence policy and the land use planning system which still allows inappropriate urban and industrial development on floodplains. The tendency has been to adopt narrowly-focused solutions rather than broader more integrated policies.[14] In flood protection cost has to be balanced against benefits arising from the avoidance of flood damage. Urban stormwater drainage systems are often designed to cope with a level of flooding which is likely to occur once in 30 years or so. It would be perfectly possible – but much more expensive – to design them to be proof against a once in a century flood.[15] In the Netherlands flood protection, against both sea and river, is taken more seriously than in Britain. Major defences are designed to be proof against once in 10,000 year events. The heavy death toll there following the North Sea storm of 1953 was the trigger for action and it has even been suggested that none of the recent floods in Britain have been quite disastrous enough to shock both government and taxpayers in a similar way. The flood defence budget is scheduled to rise from £600 million per year to £800 million – but only in 2010–11 – while the Environment Agency claim that they need £1 billion.

It is worth noting that the floods in Hull in summer 2007 were not caused by rivers bursting their banks but by pluvial floods; by the volume of rainfall exceeding the ability of pumps to remove it in a city which mostly lies below sea level. The emergency services are less able to predict, warn and cope with floods of this kind.

Because of the high levels of owner occupation, and because most owner-occupiers are insured, floods in Britain tend to be particularly expensive when measured by the total size of insurance claims. Writing in December 2007 it looks as though the hand of the government may be forced by the private sector. In the UK, unlike other European countries, protection against the cost of flooding is a standard part of property insurance. Payouts on claims from the 2007 floods absorbed about 30–40 per cent of that year's income in premiums, something which insurance companies can afford occasionally but not on a regular basis. The Association of British Insurers has asked the government to set out a 25–year strategy for flooding, and to regard recent events as a wake-up call, giving the Environment Agency more responsibility and more money, otherwise they may not be able to guarantee that they can continue to provide cover in flood-prone areas.[16] Floods, in the north west and elsewhere in Britain are, and are likely to remain, a controversial topic.

Notes

[1]   A. J. Maddock, The River Alt Level 1587–1779, Diploma in Local History Dissertation, University of Liverpool, 1995.

[2]   D. Hunt, *A History of Walton le Dale and Bamber Bridge* (Carnegie, 1997).

[3]   Hunt, *Walton le Dale*, p. 49.

[4]   Hunt, *Walton le Dale*, p. 49.

[5]   *Lancaster Gazette* 20 Dec. 1856.

[6]   *Cumberland & Westmorland Herald* 3 Nov. 1995.

[7]   *Westmorland Gazette* 12 Feb. 1831.

[8]   CRO (K), WD/PW A20.

[9]   CRO (K), WD/PW A20.

[10]   CRO (K), WD/PW A20.

[11]   Environment Agency, N.W. Region. *Lower Alt Feasibility and PAR EIA Scoping Report* (2002).

[12]   Environment Agency, *Burnley, Nelson and Colne Flood Risk Management Strategy Scoping Report* (2004).

[13]   Association of British Insurers, *Safe as Houses: Flood Risk and Sustainable Communities* (2005).

[14]   D. J. Parker, *Floods* (Routledge, 2000).

[15]   K. Smith & R. Ward, *Floods. Physical Processes and Human Impacts* (Wiley, 1998).

[16]   Association of British Insurers, *Summer Floods 2007: Learning the Lessons* (2007).

Conclusion

In this book we have tried to show that floods in north west England are in no way a recent phenomenon. Major floods are on record as far back as the late sixteenth century. Our lack of knowledge of earlier events is almost certainly due to the scarcity of records rather than a lack of floods. Many of the worst floods on record occurred at much too early a date to have anything to do with modern human-induced global warming. The weather conditions which produce floods in the north west are very complex and difficult to predict even with modern technology. The floods which devastated Carlisle in January 2005 fitted some predicted patterns for a warmer England with increased winter rainfall. However, the widespread floods which affected Central England and Yorkshire in the summer of 2007, as many commentators have noted, did not. This is not to say that global warming is not occurring and that in a warmer world flooding may not increase in both freqency and intensity. But it does indicate that global warming is not the only influence at work. The search for patterns in a complex variable like rainfall is difficult but we have seen that there were certainly distinct patterns of past flooding with decades that were relatively flood-rich and ones which were flood-poor.

There is, however, one feature of flooding about which we can be more certain. Human influences on the landscapes of north west England have become much more significant over the last four centuries. This applies to the way in which people have used the countryside and particularly how towns have been developed. Measures have been taken to provide protection against flooding but many of these would have been unnecessary if we had not undertaken so much building on vulnerable floodplains, obstructed river flows, and converted so much land in urban areas to impermeable surfaces. Global warming is an easy scapegoat (though we should not forget that in itself it is probably the result of human agency). In fact much of the blame lies nearer to home in planning decisions which should not have been taken and developments which should have been better thought through, whether paving over our front drives or being more aware of potential environmental hazards. How often, when house hunting, do people seriously investigate the flood risk for the properties in which they are interested? Should flooding histories form part of the new Home

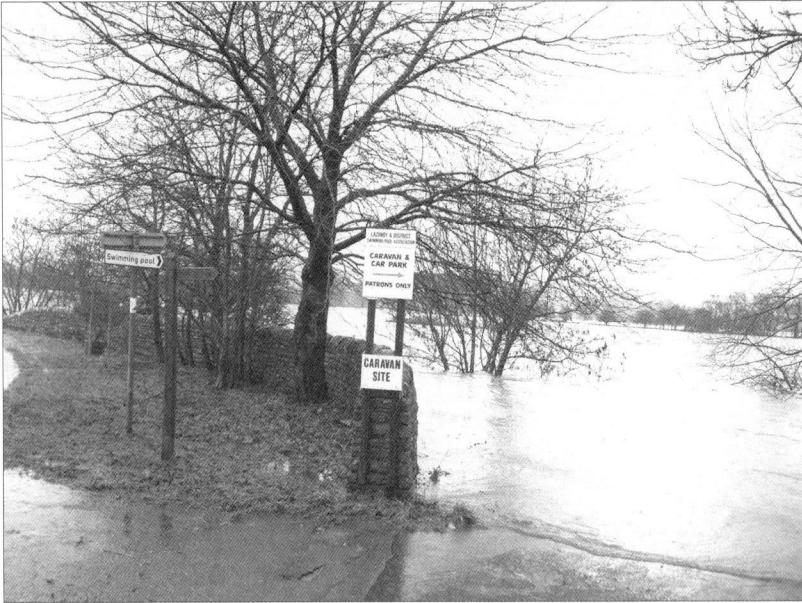

62: Flooding at Lazonby in 2005 with some ironic messages on the notice boards. Reproduced by permission of Geoff Wilson.

Information Packs? The Environment Agency has flood risk maps on the internet for potential house buyers to consult but how many actually use them?[1] Much of what happens in terms of flooding today is the result of what we have done to our river catchments in the past.

In many upland areas of Britain, it has been claimed, there has been a marked drop in the numbers of extreme floods since the 1970s, contrary to popular perceptions.[2] There has certainly been a trend to increasing winter rainfall but reduced summer rain within the region, associated with changes in the North Atlantic Oscillation. At a national scale there has also been a trend towards the north and west of the country becoming wetter overall and the south and east drier, as evidenced by the number of recent hosepipe bans.[3] In the past although there have been periods which have been flood rich and flood poor there are no clear underlying long-term trends.[4] Recent land use practices may have increased the sensitivity of north west river catchments to climatic variability.[5] We have seen that the history of flooding in north west England is complex and often dramatic. Despite modern technology for flood prediction, warning and protection, major disasters still occur. In many ways, with our comfortable twenty-first century lifestyle we are more susceptible to the risk of flooding then ever before.

There is plenty of scope for further work on the history of floods and their impact in north west England. Local studies using newspapers, documentary sources, maps and oral evidence can add a great deal to the bare outlines given here. Much of the relevant material lies buried deep in council minutes and other archive sources but there are plenty of stories about flooding still to be told.

But after everything that we have said about floods as a hazard and a danger we should remember that they are a danger because we make them so (Figure 62). Floods are a natural part of environmental systems and there are few more impressive and awesome sights than a river in full flood – viewed at a safe distance! As Gerard Manley Hopkins wrote in his poem *Inversnaid*:

What would the world be, once bereft,
Of wet and wildness? Let them be left,
O let them be left, wildness and wet;
Long live the weeds and the wilderness yet.

Notes

[1]   http://www.environment-agency.gov.uk/subjects/flood/826674/829803/

[2]   M. G. Macklin & B. J. Rumsby, 'Changing climate and extreme floods in the British uplands', *Transactions Institute of British Geographers* (New Series) 32 (2) (2007), p. 179.

[3]   J. Mayes, 'Changing regional climatic gradients in the UK', *Geographical Journal* 166 (2) (2000), pp. 125–38.

[4]   R. L. Wilby, 'When and where might climate change be detectable in UK river flows?', *Geophysical Research Letters* 33 (2006), L19407.

[5]   H. G. Orr & P. Carling, 'Hydro-climatic and land use changes in the River Lune catchment, North West England. Implications for catchment management', *River Research and Applications* 22 (2006), pp. 239–55.

Glossary

Aquifer	A rock which will absorb and hold water below ground level.
Bankfull	The discharge of a river which is just contained within its banks without flooding.
Bog burst	When a peat bog becomes so saturated with water that it loses cohesion and bursts.
Boulder bar	A deposit of large boulders left behind where a flooded river has had its flow suddenly checked.
Catchment	The area drained by a river and its tributaries.
Discharge	The quantity of water passing through a cross section of a stream or river in a unit of time, usually measured in cubic metres per second (Cumecs).
Drainage basin	See 'catchment'.
Environment Agency	The organisation responsible for implementing the government's environmental responsibilities, including flood defences.
Flood hydrograph	The graph of successive discharge measurements of a stream from the start of a flood, through its peak, to its decline.
Floodplain	The flat land on either side of a river which is periodically covered by floodwater.
Groundwater	Any water found below the surface of the earth.
Hydrograph	The graph of the discharge of a river over a period of time measured at a recording station.
Hydrology	The study of water on the earth's surface, particularly in rivers and streams.
Little Ice Age	A period from the fourteenth to the mid nineteenth century characterised by phases of colder and

sometimes wetter conditions than medieval or modern times.

Long profile	A section of the course of a stream from source to sea showing changes in gradient.
Meander	A winding curve in the course of a river.
North Atlantic Oscillation	A fluctuation of atmospheric pressure patterns in the North Atlantic area on a scale of months to decades which has a strong influence on patterns of flooding in NW England.
Overland flow	Water flowing over the surface of the ground because the underlying soil or rock is saturated.
Palaeohydrology	The study of the behaviour of rivers in the past.
Rainfall intensity	The rate at which rain falls, in millimetres or inches per hour.
Return period	The average length of time between floods of a given magnitude.
Throughflow	The movement of eater downslope through a soil rather than vertically.
Watersheds	The boundary between two catchments or drainage basins.

Bibliography

W. A. Abram, *History of Blackburn* (Toulmin, 1877).

Anon., *Lancaster Records or Leaves from Local History 1801–50* (Lancaster 1868).

Anon., *The Westmorland Book*, vol 1 (Titus Wilson, 1888–9).

A. B. Appleby, *Famine in Tudor and Stuart England* (University of Liverpool Press, 1978).

D. Archer, *Land of Singing Waters. Rivers and Great Floods of Northumbria* (Spedden Press, 1992).

D. R. Archer, F. Leesch & K. Harwood, 'Assessment of severity of the extreme River Tyne flood in January 2005 using gauged and historical information', *Hydrological Sciences Journal* 52 (5) (2007), pp. 992–1003.

W. Ashton, *The Battle of Land and Sea on the Lancashire, Cheshire and North Wales Coasts* (Heywood, 1909).

Association of British Insurers, *Safe as Houses: Flood Risk and Sustainable Communities* (2005).

Association of British Insurers, *Summer Floods 2007: Learning the Lessons* (2007).

M. A. Atkin, 'Land use and management in the upland demesne of the De Lacy estate of Blackburnshire c. 1300', *Agricultural History Review* 42 (1) (1994), pp. 1–19.

G. L. Banks, *The Manchester Man* (Gollancz, 1970).

P. A. Barker, R. L. Wilby & J. Borrows, 'A 200–year precipitation index for the central English Lake District', *Hydrological Sciences Journal* 49 (5) (2004), pp. 769–85.

A. C. Bayliss & D. W. Reed, 'The use of historical data in flood frequency estimation', *Centre for Ecology and Hydrology*, (2001).

J. Beck, 'The church brief for the inundation of the Lancashire coast in 1720', *Transactions Historic Society of Lancashire and Cheshire* 105 (1953), pp. 91–105.

N. L. Betts, 'The Antrim floods of October 1990', *Irish Geography* 25 (1992), pp. 138–45.

P. Bicknell (ed.), *The Illustrated Wordsworth's Guide to the Lakes* (Select Editions, 1984).

R. Brazdil, G. Rudiger, C. Pfister, P. Dobrovolny, J-M, Antoine, M. Barriendos, D. Camuffo, M. Deutsch, E. Guidobone, O. Kotyza & F.

Rodrigo, 'Flood events of selected European rivers in the sixteenth century', *Climatic Change* 43 (1999), pp. 239–85.

P. A. Carling, 'The Noon Hill flash floods July 17 1983. Hydrological and geomorphic aspects of a major formative event in an upland landscape', *Transactions, Institute of British Geographers* New Series 11 (1986), pp. 105–18.

P. Carling & K. Beven, 'The hydrology and geomorphological implications of floods: an overview' in P. Carling & K. Beven, (eds.), *Floods. Hydrological, Sedimentological and Geomorphological Implication* (Wiley, 1980), pp. 1–9.

P. Carling & M. S. Glaister 'Reconstruction of a flood resulting from a moraine dam failure using geomorphological evidence and dam break modelling', in L. Mayer & D. Nash (eds.), *Catastrophic Floods* (Allen & Unwin, 1987).

J. Corbett, *The River Irwell* (Heywood, 1907).

H. S. Cowper, *The oldest register of the parish of Hawkshead in Lancashire 1568–1704* (Bemrose, 1897).

H. J. Crofton, 'How Chat Moss broke out in 1526', *Transactions of the Lancashire and Cheshire Antiquarian Society* 20 (1902), pp. 139–45.

J. F. Curwen, *Kirkbie Kendall.* (Titus Wilson, 1900).

J. F. Curwen, *The Later Records of North Westmorland or the Barony of Appleby* (Titus Wilson, 1932).

T. Davie, *Fundamentals of Hydrology* (Routledge, 2003).

S. Denyer, *Traditional Buildings and Life in the Lake District* (Gollancz, 1991).

R. Doe, *Extreme Floods: a History in a Changing Climate* (Sutton, 2006).

I. Douglas, 'Geomorphology and urban development in the Manchester area, in R. H. Johnson (ed.), *The Geomorphology of North West England* (Manchester University Press, 1985), pp. 337–52.

I. Douglas, 'Urban flood plains and slopes; the human impact on the environment in the built up area', in A. Gardiner, P. Hindle, J. McKendrick & C. Perkins (eds.), *Exploring Greater Manchester. A Fieldwork Guide* (Manchester Geographical Society, 1999).

S. Downward & J. Skinner, 'Working rivers; the geomorphological legacy of English freshwater mills', *Area* 37 (2) (2005), pp. 138–47.

F. Engels, *The Condition of the Working Class in England* (Penguin 1987 ed.).

Environment Agency, *Burnley, Nelson and Colne Flood Risk Management Strategy Scoping Report* (2004).

Environment Agency, N.W. Region, *Lower Alt Feasibility and PAR EIA Scoping Report* (2002).

Environment Agency, *The Eden Catchment Flood Management Plan. Draft Scoping Report* (2005).

Environment Agency, *River Douglas Catchment Flood Management Plan. Draft Scoping Document* (2005).

Environment Agency, *River Ribble Catchment Flood Management Plan. Scoping Report* (2005).

C. W. Farrer, *A History of the Parish of North Meols* (Henry Young and Son, 1903).

F. Furedi, 'The changing meaning of disaster', *Area*, 39 (4) (2007), pp. 482–9.

E. Garnett, *The Wray Flood of 1967* (Centre for North-West Regional Studies, Lancaster University, 2002).

W. Gilpin, *Observations on the Mountains and Lakes of Cumberland and Westmorland*, vol II (1786).

E. Gorham. 'Some early ideas concerning the nature, origin and development of peat lands', *Journal of Ecology* 41 (2) (1953), pp. 257–74.

W. G. Hale & A. Coney, *Martin Mere. Lancashire's Lost Lake* (Liverpool University Press, 2005).

A. D. M. Harvey, 'The river systems of North West England in J. J. Johnston, *The Geomorphology of North West England* (Manchester University Press, 1985), pp. 122–42.

A. D. M. Harvey, 'Geomorphic effects of a 100–year storm in the Howgill Fells, North West England', *Zeitschrift fur Geomorphologie* 30 (1986), pp. 71–91.

A. D. M. Harvey, R. W. Alexander & P. A. James, 'Lichens, soil development and the age of Holocene valley floor landforms, Howgill Fells, Cumbria', *Geografiska Annaler* 66A (1984), pp. 353–66.

A. D. M. Harvey & R. C. Chiverrell, 'Carlingill, Howgill Fells' in R. C. Chiverrell, A. J. Plater & S. P. Thomas (eds.), *Quaternary of the Isle of Man and North West of England: Field Guide* (Quaternary Research Association, 2004), pp. 177–93.

D. Higgitt & E. M. Lee, *Geomorphological Processes and Landscape Change. Britain in the last 1000 years* (Blackwell, 2001).

B. P. Hindle, *Roads and Trackways of the Lake District* (Moorland, 1984).

C. Holme, *The Lonely Plough* (Oxford University Press, 1931 ed.).

G. M. Howe, H. O. Slaymaker & D. M. Harding, 'Some aspects of the flood hydrology of the upper catchments of the Severn and the Wye', *Transactions Institute of British Geographers* 41 (1967), pp. 33–58.

D. Hunt, *A History of Walton le Dale and Bamber Bridge* (Carnegie, 1997).

W. Hutchinson, *The History of the County of Cumberland and some places Adjacent*, vol I (1794).

R. M. Johnson & J. Warburton, 'Flooding and geomorphic impacts in a mountain torrent: Raise Beck, Central Lake District, England', *Earth Surface Processes and Landforms* 27 (2002), pp. 945–69.

L. J. McEwen, 'The establishment of a historical flood chronology for the River Tweed catchment, Berwickshire, Scotland', *Scottish Geographical Magazine* 106, (1990), pp. 37–48.

L. J. McEwen, 'The use of long-term rainfall records for augmenting historic flood series: a case study from the upper Dee, Aberdeenshire', *Transactions Royal Society of Edinburgh, Earth Sciences* 78 (1987), pp. 279–85.

L. J. McEwen, 'Sources for the establishment of historic flood chronologies (pre-1970) within Scottish river catchments', *Scottish Geographical Magazine* 103 (1987), pp. 132–40.

G. Maas, *A Hydrological and Hydraulic Assessment of Flooding on the River Lune at Halton,* Report Commissioned by the Halton Residents' Group (Lancaster, 2005).

N. Macdonald, A. Werrity, A. R. Black & L. J. McEwen, 'Historical and pooled flood frequency analysis for the River Tay at Perth, Scotland', *Area,* 38 (1) (2006), pp. 34–46.

M. G. Macklin, D. G. Passmore & B. T Rumsby, 'Climatic and cultural signals in Holocene alluvial sequences: the Tyne basin, northern England' in S. Needham & M. G. Macklin (eds.), *Alluvial Archaeology in Britain* (Oxbow, 1992), pp. 123–39.

M. G. Macklin & B. J. Rumsby, 'Changing climate and extreme floods in the British uplands', *Transactions Institute of British Geographers* (New Series) 32 (2), (2007), pp. 168–86.

M. G. Macklin, B. T. Rumsby & J. Heap, 'Flood alleviation and entrenchment. Holocene valley flood development and transformation in the British uplands', *Geological Society of America Bulletin* 104 (1992), pp. 631–43.

A. J. Maddock, The River Alt Level 1587–1779, Unpub. Diploma in Local History Dissertation, University of Liverpool (1995).

G. Manley, *Climate and the British Scene* (Collins, 1962).

L. Markham, *The Lancashire Weather Book,* (Countryside Books, 1995).

T. J. Marsh, Sand pilots: a study of the history and chronology of the guides to Morecambe Bay Sands 1501–2006, Unpub. MA Dissertation, Lancaster University D/5912 (2006).

J. Mayes, 'Changing regional climatic gradients in the UK', *Geographical Journal* 16 (2) (2000), pp. 125–38.

R. Millward, *Lancashire* (Hodder, 1955).

S. Murphy, *Grey Gold. Men, Mining and Metallurgy at the Greenside Lead Mines in Cumbria, England 1825–1962* (Moiety, 1996).

H. G. Orr, The Impact of Recent Changes in Land Use and Climate on the River Lune: Implications for Catchment Management. Unpub. PhD Thesis, Lancaster University (2000).

H. G. Orr & P. Carling, 'Hydro-climatic and land use changes in the River Lune catchment, North West England. Implications for catchment management', *River Research and Applications* 22 (2006), pp. 239–55.

S. Owen (ed.), *Rivers and the British Landscape* (Carnegie, 2005).

D. H. Parker, *Floods* vol 1 (London, 2000).

D. G. Passmore, M. G. Macklin, A. G. Stevenson, C. F. O'Brien & B. A. Davies, 'A Holocene alluvial sequence in the lower Tyne valley, northern Britain. A record of river response to environmental change', *The Holocene* 2 (7) (1992), pp. 138–47.

E. C. Penning-Rowsell & J. W. Handmer, 'Flood hazard management in Britain: a changing scene', *Geographical Journal* 152 (2) (1988), pp. 209–20.

H. R. Potter, 'The use of historic records for the augmentation of hydrological data', *Institute of Hydrology Report* no.46 (1978).

D. B. Prior & N. L. Betts, 'Flooding in Belfast', *Irish Geography* 7 (1974), pp. 1–18.

D. W. Proctor, *Memorials of Manchester Streets* (Sutcliffe, 1874).

M. Robinson & K. J. Beven, 'The effect of mole drainage on the hydrological response of a swelling clay soil', *Journal of Hydrology* 64 (1983), pp. 205–23.

A. Robson, 'Evidence for trends in UK flooding', *Phil. Trans. Roy. Soc. London,* A360 (2002), pp. 1327–53.

J. E. Roe, & A. Parker, 'Techniques for validating the historic record of lake sediments. A demonstration of their use in the English Lake District', *Applied Geochemistry* 11 (1996), pp. 211–15.

W. Rollinson, 'Schemes for the reclamation of land from the sea in North Lancashire during the eighteenth and nineteenth centuries', *Transactions of the Historic Society of Lancashire and Cheshire* 115 (1963), pp. 133–45.

A. W. Rumney (ed.), *Tom Rumney of Mellfell (1764–1835) by Himself as set out in his Letters and Diary* (Titus Wilson, 1936).

B. T. Rumsby & M. G. Macklin, 'Channel and floodplain response to recent abrupt climatic change. The Tyne basin, northern England', *Earth Surface Processes and Landforms* 19 (1994), pp. 499–515.

G. Smith, 'Dreadful storm in Cumberland', *Gentleman's Magazine* 24 (1754), pp. 476–7;

K. Smith & G. Tobin, *Human Adjustment to the Flood Hazard* (Longman, 1979).

K. Smith & R. Ward, *Floods. Physical Processes and Human Impacts* (Wiley, 1998).

J. Somervell, *Water-powered mills of South Westmorland* (Titus Wilson, 1930).

I. Tyler, *Greenside and the Mines of the Ullswater Valley* (Bluerock, 2001).

J. G. Tyrrel & K. J. Hickey, 'A flood chronology for Cork city and its climatological background' *Irish Geography* 24 (1991), pp. 81–90.

G. Walker, J. Fairbairn, G. Smith & G. Mitchell, *Environmental Quality and Social Deprivation, Phase II. National Analysis of IPCC, Flood Hazard and Air Quality* (Environment Agency, 2003).

R. P. D. Walsh, R. N. Hudson & K. A. Howells, 'Changes in the magnitude and frequency of flooding and heavy rainfall in the Swansea valley since 1875', *Cambria* 9 (1983), pp. 36–60.

R. Watson & M. M. McClintock, *Traditional Houses of the Fylde* (Centre for North-West Regional Studies, Lancaster University, 1979).

I. D. Whyte, *Transforming Fell and Valley. Landscape and Parliamentary Enclosure in North West England* (Centre for North-West Regional Studies, Lancaster University, 2003).

A. J. L. Winchester (ed.), *The Diary of Isaac Fletcher of Underwood, Cumberland, 1756–178* (Cumberland and Westmorland Antiquarian and Archaeological Society extra series, 1994).

A. J. L. Winchester & A. G. Crosby, *England's Landscape. The North West* (English Heritage, 2006).

C. W. J. Withers & L. J. McEwen, 'Historical records and geomorphological events: the Solway Moss 'eruption' of 1771', *Scottish Geographical Magazine* 105 (3) (1989), pp. 149–57.

Wyre Borough Council. *Policy Statement on Flood and Coastal Defence* (2000).

Index